BIOHALOGENATION:
Principles, Basic Roles and Applications

BIOHALOGENATION:

Principles, Basic Roles and Applications

SAUL L. NEIDLEMAN, PhD
Distinguished Research Fellow
Cetus Corporation, Emeryville, California
and
J. GEIGERT, PhD
Director, Technical Development
Cetus Corporation, Emeryville, California

ELLIS HORWOOD LIMITED
Publishers · Chichester

Halsted Press: a division of
JOHN WILEY & SONS
New York · Chichester · Brisbane · Toronto

First published in 1986 by
ELLIS HORWOOD LIMITED
Market Cross House, Cooper Street, Chichester, West Sussex, PO19 1EB, England

The publisher's colophon is reproduced from James Gillison's drawing of the ancient Market Cross, Chichester.

Distributors:
Australia and New Zealand:
Jacaranda-Wiley Ltd., Jacaranda Press,
JOHN WILEY & SONS INC.
GPO Box 859, Brisbane, Queensland 4001, Australia
Canada:
JOHN WILEY & SONS CANADA LIMITED
22 Worcester Road, Rexdale, Ontario, Canada
Europe and Africa:
JOHN WILEY & SONS LIMITED
Baffins Lane, Chichester, West Sussex, England
North and South America and the rest of the world:
Halsted Press: a division of
JOHN WILEY & SONS
605 Third Avenue, New York, NY 10158, USA

British Library Cataloguing in Publication Data
Neidleman, Saul L.
Biohalogenation: principles, basic roles and applications. —
(Ellis Horwood series in organic chemistry)
1. Halogenation 2. Biological chemistry
I. Title II. Geigert, J.
574.19'283 QP517.H3

ISBN 0–85312–984–3 (Ellis Horwood Limited)
ISBN 0–470–20285–8 (Halsted Press)

Typeset by Ellis Horwood Limited
Printed in Great Britain by Butler & Tanner, Frome, Somerset

Table of Contents

Acknowledgements

The authors wish to thank the many friends and colleagues who made it possible for us to write this book. Dr J. C. Hunter (microbial ecology), Dr T.-N. Liu (protein chemistry), and the late Dr J. E. Harrison (mechanisms) are especially mentioned for their helpful discussions during the early stages of formulating the manuscript. We thank Dr D. J. Dalietos and S. K. DeWitt for their invaluable contributions to our study of many of the haloperoxidase reactions reported here. We thank E. Jarvis for the many tedious hours she spent in typing the manuscript. The authors thank Cetus Corporation for allowing us the time and help to prepare this work. Finally, we would like to thank the enzymes, substrates, products, and reaction conditions without which there would be no book.

1

Introduction

The understanding of the significance of biological halogenation in nature is still at a primitive level of development when compared to that of biological phosphorylation, oxidation, and reduction, for example. The understanding of the applications of biological halogenation in nature is also at a primitive level when compared to that of the organic chemistry industry, where halogenation reactions have achieved primary importance as a means of generating activated intermediates in many synthetic sequences and also as an integral part of preparing biologically active substances, such as pesticides.

The purpose of this monograph on biological halogenation is to explore the known and speculate on the unknown in regard to the basic roles and applications of biohalogenation. A brief review of the topics to be discussed will serve to indicate the scope of this work and to indicate the logical progression the thoughts will take.

Nature has synthesized a dazzling array of halogenated natural products, referred to as halometabolites, broadly differing in their chemical structures and properties. The very diversity of these compounds, both in their chemistry and their sources, suggests that halometabolites have biochemical importance to living creatures far beyond that which our present, sketchy understanding offers. A selection of examples from the wide spectrum of halogenated compounds from many natural sources will be considered in Chapter 2.

In concert with this breadth of chemical diversity in halometabolites and in response to the variety of conditions under which they are synthesized, nature has created a range of biocatalysts, referred to as enzymes, to favor their biosynthesis. Given the variability in the environmental pressures that living creatures endure, sophisticated and subtle catalyst design for process optimization has resulted. Chapter 3 will be devoted to describing the halogenating enzymes, usually referred to as haloperoxidases, their sources, and the many reactions they have been demonstrated to catalyze in laboratory studies. Chapter 4 will consider their properties and the mechanism of the reactions catalysed by the halogenating enzymes.

The marine environment is the richest source of halometabolites and halogenating enzymes. Chapter 5 will explore the possible importance of these compounds and catalysts in the function and welfare of marine organisms. The relationship of the halide ion composition of seawater and the subtleties of enzyme design in the marine biosphere will be emphasized as prime examples of biological response to environmental conditions.

That halogenating enzymes do have a basic role in biological survival will be further elaborated upon in Chapter 6. Specifically, the functioning of these enzymes and their reaction products as part of human defense mechanisms will be discussed. The growing significance of the medical implications of these enzymes will be emphasized.

In the natural progression of research, it is often the case that commercial applications of enzyme reactions are sought. Chapter 7 will summarize the present situation with regard to the production of commodity and specialty chemicals with halogenating enzymes and with regard to the use of these enzymes in diagnostics and pollution control.

Until this point, this monograph has been concerned with the incorporation of halogen into organic compounds in nature and in the laboratory. Due mainly to man's distribution of halogenated compounds at various locations on the surface of the earth, a pressure has been placed on many ecosystems to degrade such compounds. Chapter 8 concerns itself, then, with biodegradation and dehalogenation of halogenated chemicals.

The monograph will conclude in Chapter 9 with some speculations as to where studies on biological halogenations may be heading.

This monograph is not an exhaustive review of all the work done on biological halogenation. Rather, it is intended to be a broad survey of the field with the objective of stimulating others, in diverse scientific disciplines, to think about and do research in this still modest arena of study.

2

Halometabolites and their sources

INTRODUCTION

In 1896 an iodinated amino acid, subsequently identified as 3,5-diiodotyrosine, was isolated from the coral *Gorgonia cavolinii* and became the first reported, naturally occurring, halogenated organic compound (Drechsel 1896). The collection of halometabolites has grown dramatically since that time, numbering well over 700 today. Many excellent reviews exist, including the following: 1) general review (Suida & DeBernardis 1973); 2) fungi (Turner 1971, Turner & Aldrich 1983); 3) microbial (Neidleman 1975); and 4) marine organisms (Moore 1978, Thompson 1978, Scheuer 1981, Erickson 1983, Naylor *et al.* 1983, Faulkner 1984). This chapter is not intended as a complete catalog of these halometabolites. Instead, the objective of this discussion is to indicate the impressive range of halometabolite occurrence in biological systems and to illustrate the spectacular diversity of chemical species represented.

BROMO- AND CHLOROMETABOLITES

Chlorine is by far the most abundant member of the halogen family on the earth, followed by bromine: 0.031% and 1.6×10^{-4}% by weight in rocks, and 19 mg/ml and 65 μg/ml by weight in seawater, respectively. In certain natural niches such as the Dead Sea and salt bitterns, these halogen concentrations can increase ten-fold or more. The versatile use of this resource by organisms to produce bromo- and chlorometabolites is exciting to unfold, as illustrated by the following:

1. **Bacteria** (Fig. 2.1)

Some commercially important chlorinated antibiotics are produced by terrestial bacteria, for example, chlortetracycline and chloramphenicol. In these and other cases, as a result of replacing chloride with bromide in the fermentation medium, the bacteria can produce brominated analogs (Suida & DeBernardis 1973, Neidleman 1975). This will be illustrated in Chapter 3. Brominated metabolites are naturally produced both in marine and terrestrial bacteria. The bromopyrrole produced by *Pseudomonas bromoutilis* is an extremely bromine-rich metabolite, having greater then 70% of its weight as this element (Suida & DeBernardis 1973). Even a casual perusal of Fig. 2.1 shows that the chemical diversity of halogenated compounds is most impressive. This point will be even more evident as further figures are studied.

Figure 2.1
Bromo– and chlorometabolites of bacteria

Structure	Source	Ref.
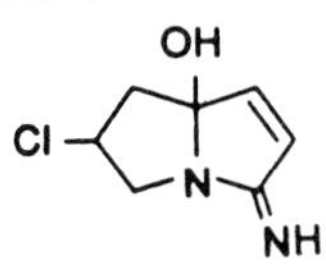 clazamycin B	*Streptomyces puniceus*	(1)
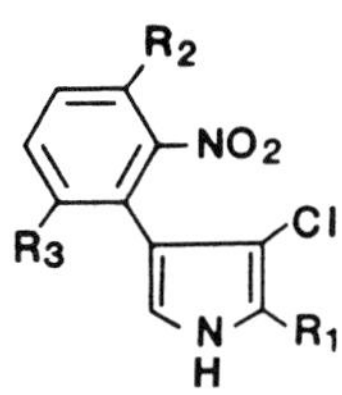		
R_1 R_2 R_3		
H H H	*Pseudomonas pyrrolnitrica*	(2)
H Cl H	*P. pyrrocinia*	(2)
Cl H H	*P. pyrrocinia*	(2)
H Cl OH	*P. pyrrocinia*	(2)
Cl Cl H	*P. fluorescens*	(2)
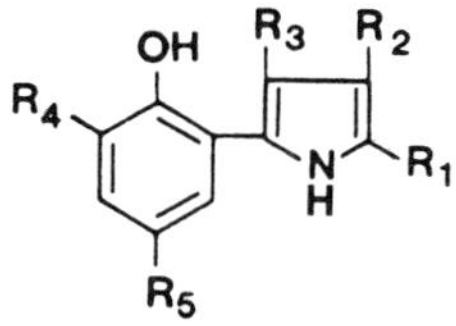		
R_1 R_2 R_3 R_4 R_5		
Cl H NO_2 Cl Cl	*Actinosporangium vitaminophilum*	(3)
Br Br Br Br Br	*Pseudomanas bromoutilis*	(2)

(*continued next page*)

Structure	Source	Ref.
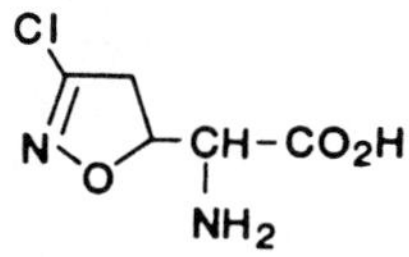	*Streptomyces sviceus*	(4)
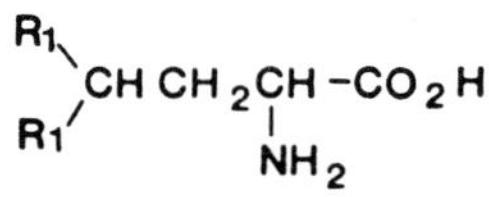		
R_1 R_2		
Cl Cl	*Streptomyces armentosus*	(4)
CH_3 Cl	*Streptomyces griseosporeus*	(6)
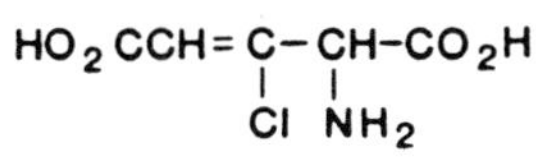	*Streptomyces viridogenes*	(7)
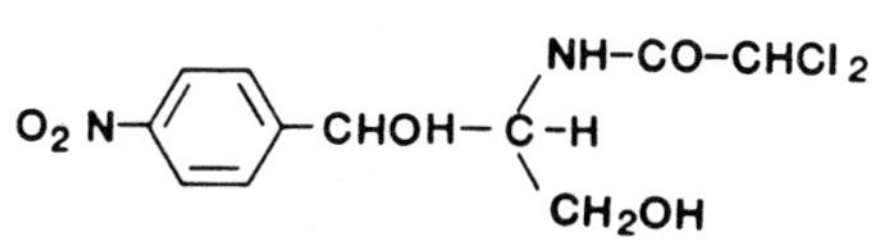 chloramphenicol	*Streptomyces venezuelae*	(4)
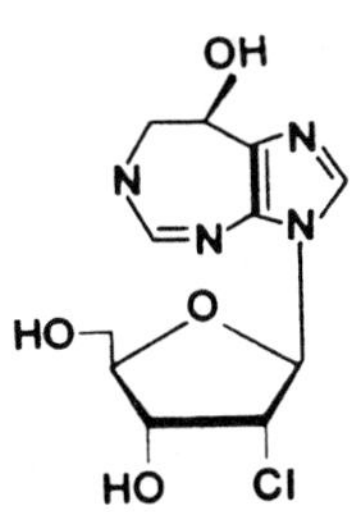 chloropentostatin	*Actinomyces* sp.	(5)
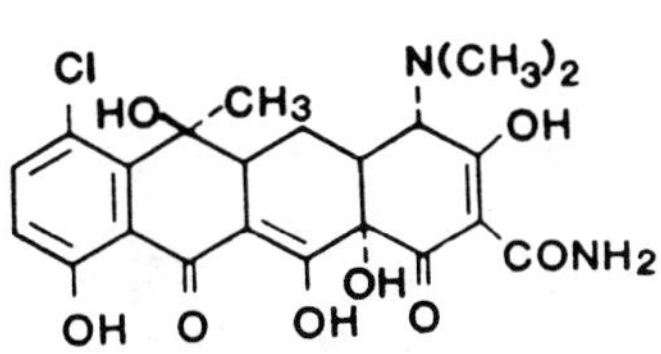 chlortetracycline	*Streptomyces aureofaciens*	(4)

(*continued next page*)

Structure	Source	Ref.
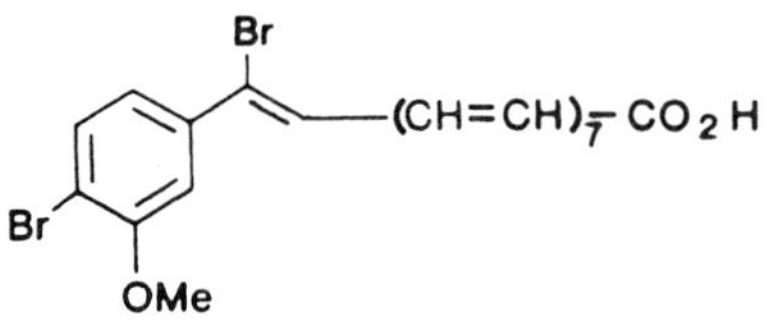	*Xanthomonas* sp.	(8)
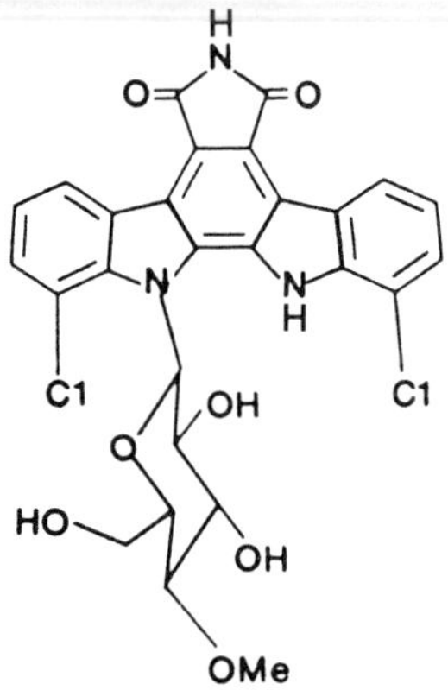 rebeccamycin	*Norcardia aerocolonigenes*	(9)

References:

1. Dolak & DeBoer (1980).
2. Suida & DeBernardis (1973).
3. Ezaki *et al.* (1983a).
4. Neidleman (1975).
5. Smal & Baker (1984).
6. Narayanan *et al.* (1980).
7. Chaiet *et al.* (1984).
8. Starr *et al.* (1977).
9. Nettleton *et al.* (1984).

2. Fungi (Fig. 2.2)

Turner (1971, 1983), in his two-volume series on fungal metabolites, lists over 70 compounds containing chlorine, several of which are shown here. Although the brominated analog of griseofulvin can be produced by substituting bromide for chloride in the fermentation medium (Suida & DeBernardis 1973), no naturally produced bromometabolites have been reported from fungi. This is surprising since many fungi have been shown to accumulate bromide, even up to a concentration of 3% of the dry weight, as was the case for the mushroom *Amanita phalloides* (Stijve 1984).

A little over two decades ago, it was believed that chlorometabolite occurrence was a rarity despite the abundance of inorganic chloride in nature (Petty 1961). Clutterbuck and co-workers (1940), surveyed 139 species or strains of fungi for their ability to convert inorganic chloride to chlorometabolites (i.e., organic chlorine). The data obtained are given in Table 2.1. From these data, Clutterbuck and co-workers concluded that extensive conversion of inorganic chloride to organic chlorine was of rare occurrence. As a gross analysis of the data (i.e., assuming that 25% conversion is the minimum limit), this is true. However, if the calculations are instead based upon the concentrations of the final products, an opposite conclusion can be drawn. It can be determined that the total chloride/ml in experimental flasks was 240 μg/ml. If all this chloride were converted to a hypothetical halometabolite with molecular weight of 355, and containing one chlorine atom, the concentration of this compound would be 2.4 mg/ml. This is a very substantial production level for a microbial metabolite. Even if only 10% of the available inorganic chloride were utilized, the resulting chlorometabilite yield, 240 μg/ml, falls within the boundaries of production for many metabolites. Product yields are often even lower. Therefore, reasonable levels of chloride incorporation into organic molecules (i.e., assuming that 5% conversion is the minimum limit) occurred in at least 50% of the fungi tested. As a matter of fact, the extreme position has been taken that almost all of the fungi in this survey formed chlorometabolites (Neidleman 1975).

Table 2.1 – Conversion of inorganic chloride to organic chlorine by 139 surveyed fungi*.

% inorganic chloride converted to organic chlorine†	Percent of surveyed fungi
>25	<4
15–25	9
10–15	17
7–10	8
6–7	10
5–6	12
3–5	11
2–3	15
1–2	10
0–1	5

* Clutterbuck *et al.* (1940).

† $\frac{\text{(organic chlorine)}}{\text{(total chloride added)}} \times 100.$

Among the fungi studied by Clutterbuck and co-workers were several species known to yield chlorometabolites, although the specific strain cannot be confirmed to do so for all cases. Some representative organisms in this group are shown in Table 2.2. Also included among the fungi studied were several species not known to yield any chlorometabolites. Some representative organisms in this group are shown in Table 2.3. Further study of members of this list of fungi might yield some previously unreported chlorometabolites.

Table 2.2 – Conversion of inorganic chloride to organic chlorine by chlorometabolite-producing fungi.

Fungus	% inorganic chloride converted to organic chlorine*	Known chlorometabolites produced
Aspergillus terreus (Strain No. 45)	97	geodin, erdin
Caldariomyces fumago (Strain Ag. 92)	96	caldariomycin
Aspergillus terreus Thom (Strain Am. 2)	34	geodin, erdin
Aspergillus terreus Thom (Strain Ab. 1)	15–25	geodin, erdin
Penicillium raistrickii G. Smith	10–15	griseofulvin
Aspergillus terreus Thom (Strain No. 3)	10–15	geodin, erdin
Aspergillus flavipes (Bainier and Sartory) Thom and Church	7–10	geodin
Aspergillus terreus Thom (Strain Ac. 107)	6–7	geodin, erdin
Aspergillus fumigatus Fresenius (Strain H.A. 1)	6–7	7-chloroemodin
Penicillium griseo-fulvum Dierckx (Strain P. 68)	5–6	griseofulvin
Aspergillus ochraceus Wilhelm	5–6	ochratoxin A
Aspergillus nidulans (Eidem) Winter	5–6	nidulin, nornidulin
Aspergillus fumigatus Fresenius (Strain No. 47)	5–6	7-chloroemodin
Penecillium raciborskii Zaleski	3–5	griseofulvin
Aspergillus candidus Link	3–5	chloroflavonin

* Clutterbuck *et al.* (1940).

Table 2.3 – Conversion of inorganic chloride to organic chlorine by fungi not reported as producers of chlorometabolites.

Fungus	% inorganic chloride converted to organic chlorine*
Absidia spinosa Lendner	27
Aspergillus niger van Tieghem (Strain Ac. 1)	15–25
Aspergillus itaconicus Kinoshita	15–25
Aspergillus versicolor (Vuillemin) Tiraboschi (Strain No. 35)	15–25
Penicillium commune Thom	15–25
Penicillium griseo-roseum Dierckx	15–25
Penicillium westlingi Zaleski	15–25
Alternaria tenuis Nees	15–25
Syncephalastrum racemosum (Cohn) Schroter	15–25
Trichoderma viride Persoon ex Fries	15–25
Cladosporium herbarum (Persoon) Link	10–15
Penicillium rugulosum Thom	10–15
Penicillium frequentans Westling (Strain P. 28)	10–15
Penicillium citrinum Thom (Strain P. 6)	10–15
Penicillium chrysogenum (Strains P. 19 and H.P. 1)	10–15
Penicillium chrysogenum Thom (Strain P. 17)	10–15

* Clutterbuck *et al.* (1940).

Figure 2.2

Chlorometabolites of fungi

Structure	Source
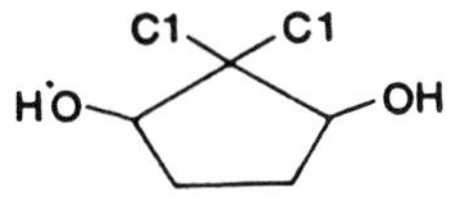 caldariomycin	*Caldariomyces fumago*
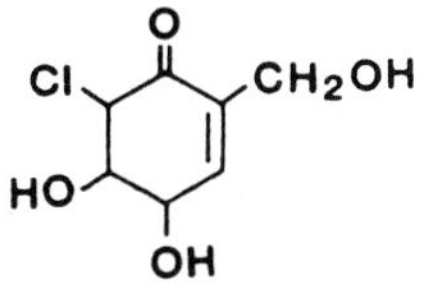	*Phyllosticta* sp.

(*continued next page*)

Structure	Source

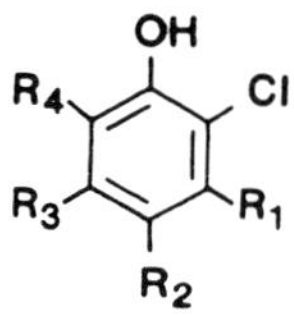

R_1	R_2	R_3	R_4	Source
H	OH	CH_2OH	H	*Penicillum martensii*
H	OH	H	CH_2OH	*Pencillium canadense*
H	CH_2CO_2H	H	H	*Marasmius palmivorus*
H	Cl	H	H	*Penicillium* sp.
Cl	OMe	Cl	Cl	*Drosophilia subatrata*

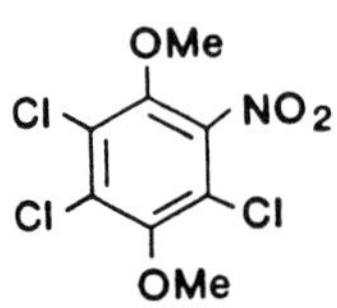

Fomes robiniae

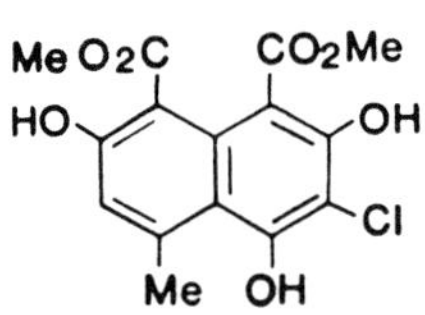

Verticillium lamellaiola

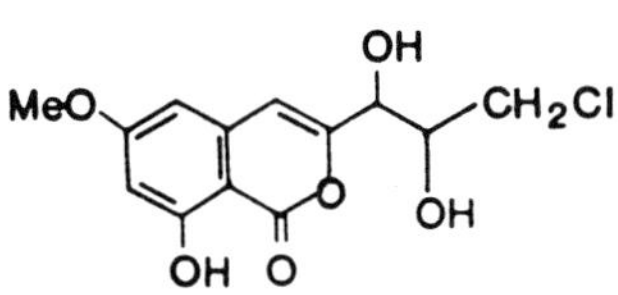

LL-Z1640-5 — unidentified fungus

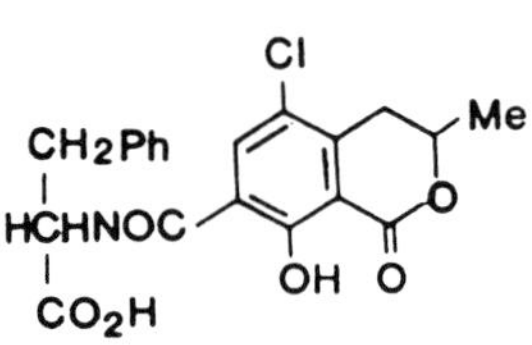

ochratoxin A — *Penicillium viridicatum*

(continued next page)

Structure	Source

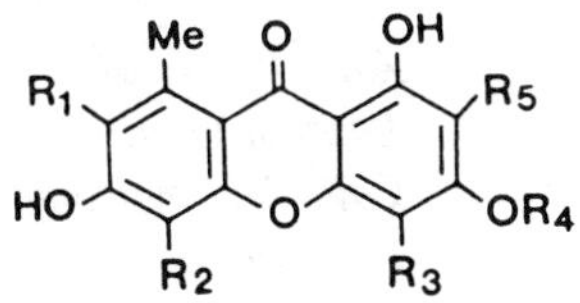

R_1	R_2	R_3	R_4	R_5	Source
H	Cl	H	H or Me	H	*Lecanora straminea*
H	H	Cl	H or Me	H	*L. straminea*
H	H	H	H	Cl	*L. straminea*
H	H	Cl	H	Cl	*L. straminea*
H	Cl	Cl	H	H	*L. straminea*
H	Cl	H	Me	Cl	*L. contractula*
Cl	H	H	H or Me	Cl	several lichens
Cl	Cl	H	H or Me	Cl	several lichens
Cl	H	Cl	H or Me	Cl	several lichens
Cl	Cl	Cl	H	Cl	several lichens

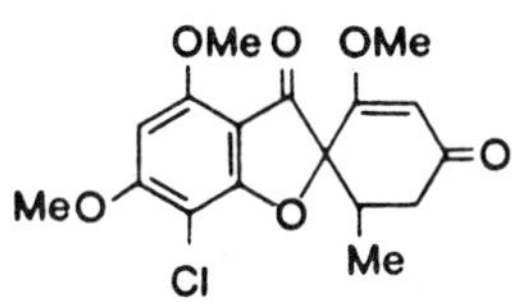

griseofulvin — *Penicillium griseofulvum*

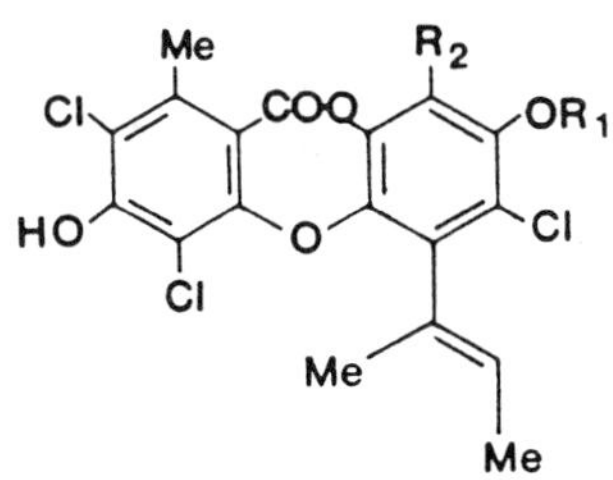

R_1	R_2		Source
Me	Me	nidulin	*Aspergillus nidulans*
H	C_4H_9	nasrin	*Aspergillus unguis*

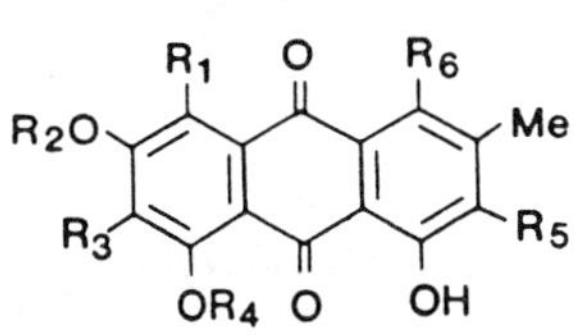

R_1	R_2	R_3	R_4	R_5	R_6	Source
H	H or Me	Cl	H	H	H	*Anaptychia obscurata*
H	H	Cl	H	H	OH	*Valaria rubricosa*

(continued next page)

Structure						Source
Cl	H	Cl	H	H	H	*Anaptychia obscurata*
OH	H	Cl	H	H	H	*Lasallia papulosa*
H	Me	Cl	Me	H	H	*Caloplaca xanthapsis*
Cl	H	H	Me	CO_2H	H	*Dermocybe sanguinea*
Cl	H	H	Me	CO_2H	OH	*D. sanguinea*

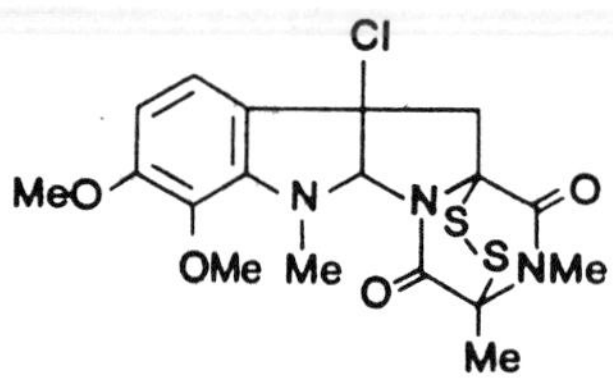

sporidesmin H — *Penicillium chartarum*

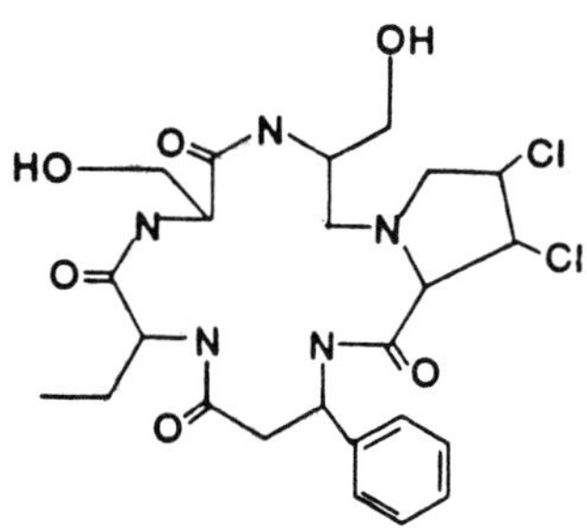

islanditoxin — *Penicillium islandicum*

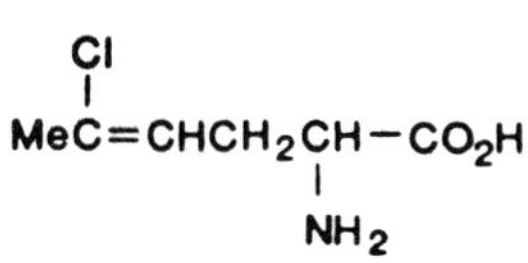

A. solitaria

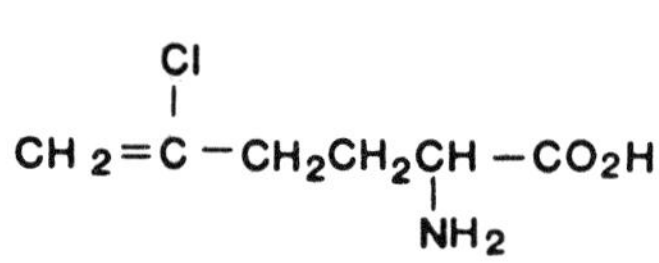

A. onusta

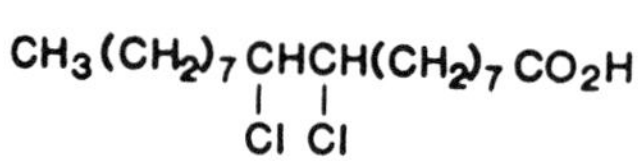

Verticillium dahliae

Reference: Turner (1971), Turner & Aldridge (1983).

3. **Algae** (Figs. 2.3, 2.4)

The algae, particularly the marine algae, are by far the richest known source of halometabolites. In a survey involving 186 algal species, the average halometabolite concentrations were 19.8 μg chlorometabolites and 19.3 μg bromometabolites per g of wet weight. On the average, about 20% of the extractable molecules in these marine algae contained halogen atoms (Hager 1982). The red algae have been extensively studied by natural product chemists, yielding an imposing variety of halometabolites, as seen in Fig. 2.3. The brown, green, and blue-green algae have also yielded a respectable array of halometabolites, as seen in Fig. 2.4. A most interesting discovery was the occurrence of the chlorinated compound, cyanobacterin, in a freshwater blue-green alga. This was the first halogenated compound ever found in any freshwater algal species (Mason *et al.* 1982).

It is in the marine algae that the chemical diversity of halogenated metabolites reaches its apex: acyclics, monoterpenes, sesquiterpenes, diterpenes, phenols, indoles, acetylenes, and other chemical species. As with certain bacteria and fungi, the alga, *Ochromonas danica*, which normally produces chlorosulfolipids, can synthesize the corresponding group of bromosulfolipids by replacing chloride with bromide in the growth medium (Haines 1973). The versatility of the algae in organic halogenation would please any practicing organic chemist. More will be said concerning the chemical and biological significance of this extraordinary activity in Chapter 5.

Figure 2.3

Bromo– and chlorometabolites of red algae

Structure	Source	Ref.
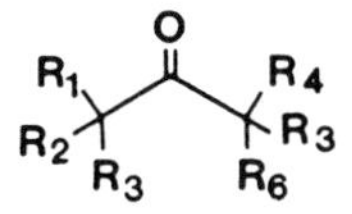	*Asparagopsis* sp.	(1)

R_1	R_2	R_3	R_4	R_5	R_6
Cl	H	H	Br	Br	H
Br	H	H	Br	Br	H
I	H	H	Br	Br	H
Br	Br	H	Br	Br	H
Br	Cl	H	Br	Br	H
Cl	Cl	H	Br	Br	H
Br	H	H	Br	Cl	H
Br	Cl	H	Br	Cl	H
Cl	Cl	H	Br	Cl	H
Br	H	H	Cl	Cl	H
Cl	H	H	Cl	Cl	H
Cl	Cl	H	Cl	Cl	H
Cl	Cl	Cl	Cl	Cl	Cl

(continued next page)

Structure	Source	Ref.
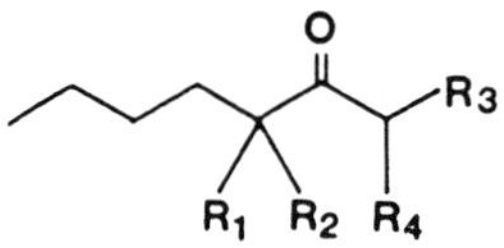	*Bonnemaisonia hamifera*	(1)
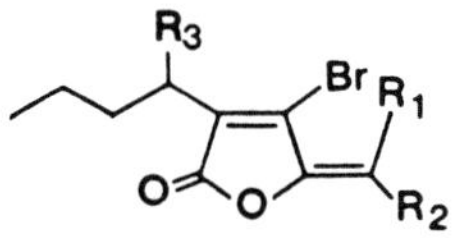	*Delisea fimbriata*	(1)
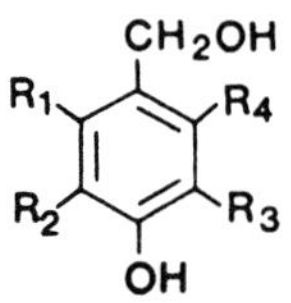	*Polysiphonia lanosa*	(2)

First structure (*Bonnemaisonia hamifera*):

R_1	R_2	R_3	R_4
Br	Br	Br	H
H	Br	Br	Br
Br	Br	I	H
Br	Br	Br	Br

Second structure (*Delisea fimbriata*):

R_1	R_2	R_3
H	Br	H
Br	H	H
Br	Br	H
H	Br	OH
H	Cl	OH
Cl	H	OH
I	H	OH
Br	Br	OH

Third structure (*Polysiphonia lanosa*):

R_1	R_2	R_3	R_4
H	OH	Br	OH
H	Br	Br	H
H	OH	Br	Br
Br	OH	Br	Br

(*continued next page*)

Structure	Source	Ref.
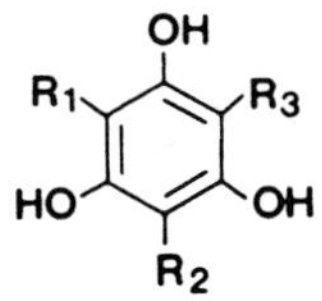	*Rhabdonia verticillata*	(3)
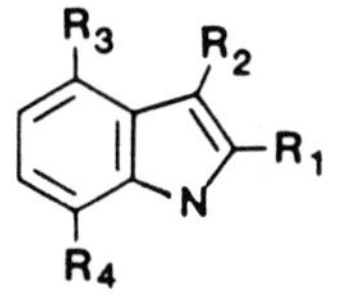	*Rhodophyllis membranacea*	(4)
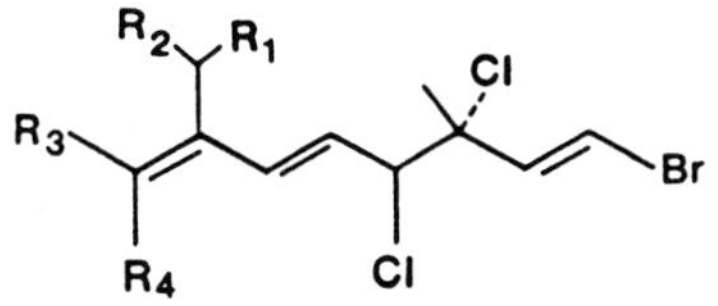	*Plocamium* sp.	(5)

R_1	R_2	R_3
H	Br	Br
H	Cl	Br
Br	Br	Br
Br	Br	Cl

R_1	R_2	R_3	R_4
Cl	Cl	H	Br
Br	Cl	H	Cl
Cl	Cl	Br	Cl
Cl	Cl	Br	Br
Br	Cl	Br	Br
Cl	Cl	Cl	Cl
Br	Br	Br	Br

R_1	R_2	R_3	R_4
H	H	H	Br
H	H	Br	Br
Cl	Cl	H	Cl
Cl	Cl	Cl	H

(continued next page)

Structure	Source	Ref.
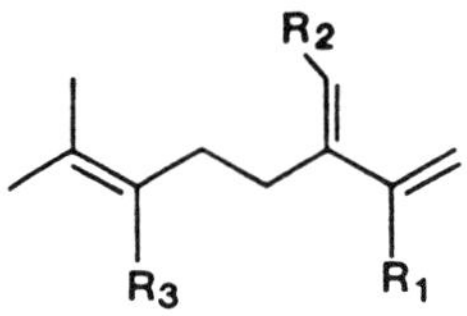	*Desmia hornemanni*	(2)
R_1 R_2 R_3 Cl H H Br H H H Br H Cl Br H Br Br Cl Br Cl Br		
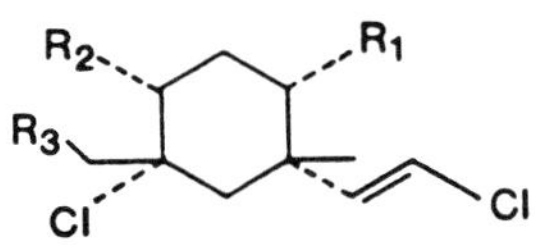	*Plocamium violaceum*	(5)
R_1 R_2 R_3 Br Br H Cl Cl Br violacene Cl Br Br		
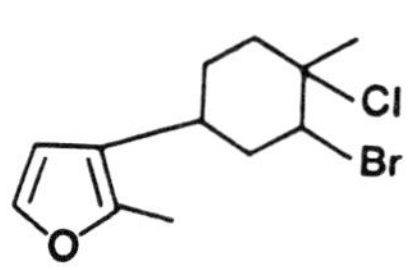 furocaespitane	*Laurencia caespitosa*	(2)
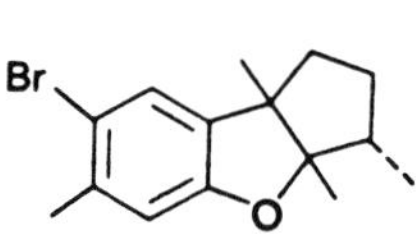 aplysin	*Laurencia okamuri*	(2)
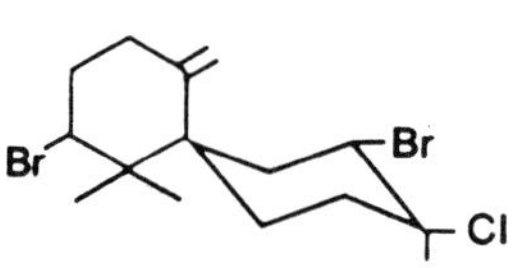 nidificene	*Laurencia nidifica*	(2)

(*continued next page*)

Structure	Source	Ref.
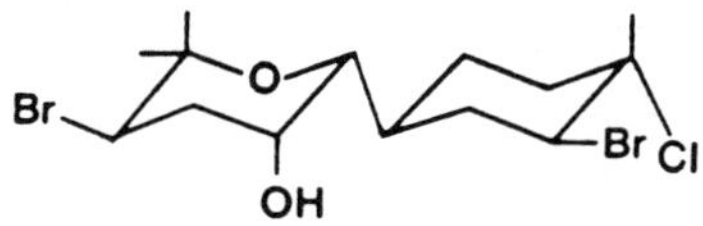 caespitol	*Laurencia caespitosa*	(2)
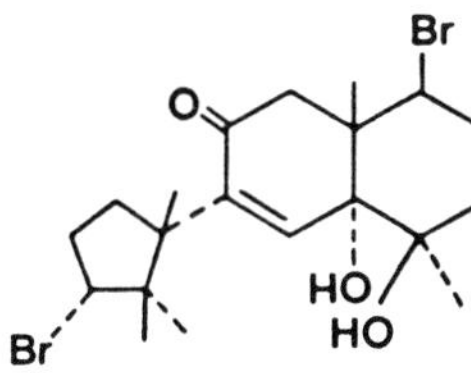 neoirieone	*Laurencia irieii*	(6)
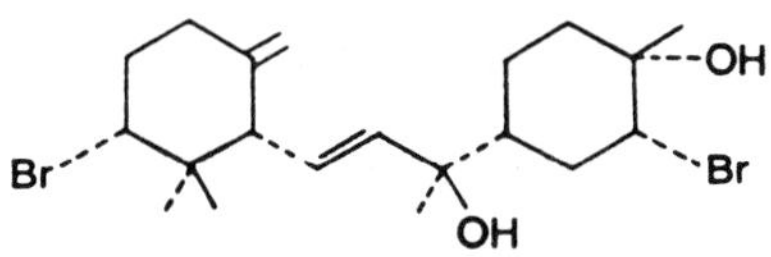 obtusadiol	*Laurencia obtusa*	(6)
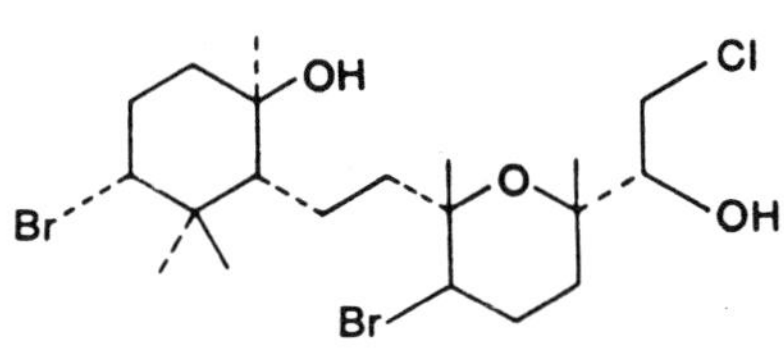 laurencianol	*Laurencia obtusa*	(6)
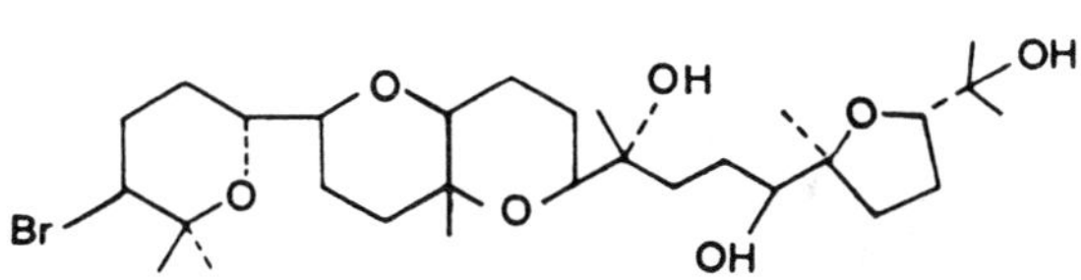 thyrsiferol	*Laurencia thyrsifera*	(6)
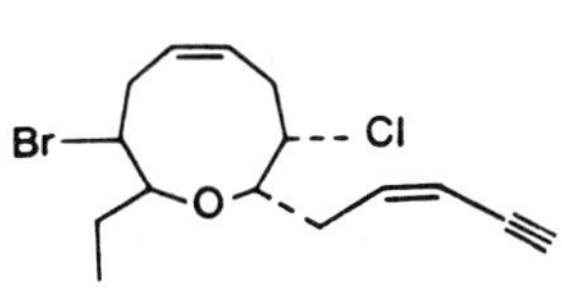 obtusenyne	*Laurencia obtusa*	(6)

(*continued next page*)

Structure	Source	Ref.
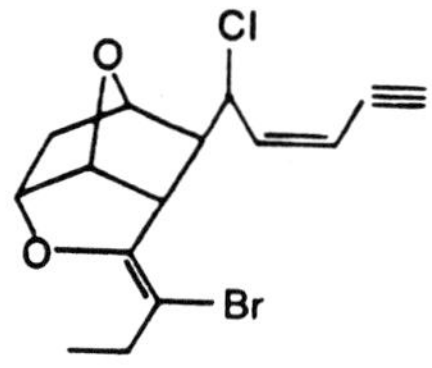 cis-maneonene A	*Laurencia nidifica*	(7)
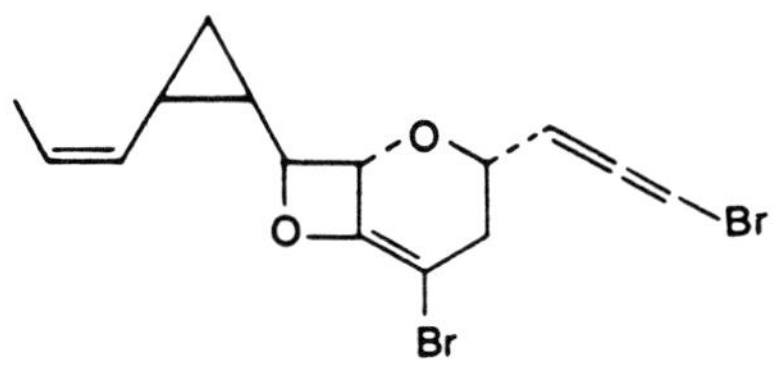 deoxyokamurallene	*Laurencia okamurai*	(6)
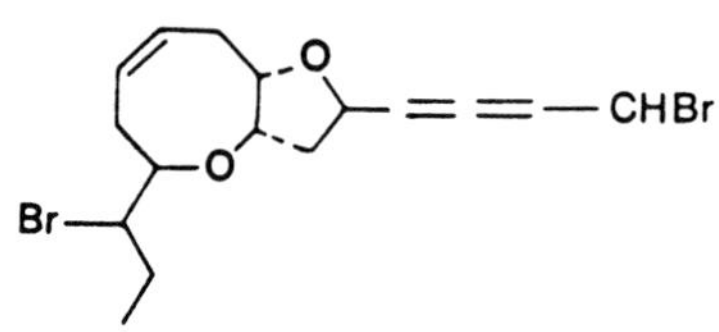 laurallene	*Laurencia nipponica*	(6)

References:

1. Moore (1978).
2. Fenical (1975).
3. Blackman & Matthews (1982).
4. Brennan & Erickson (1978).
5. Naylor *et al.* (1983).
6. Erickson (1983).
7. Ragan (1981).

Figure 2.4

Bromo– and chlorometabolites of other algae

Structure	Source	Ref.
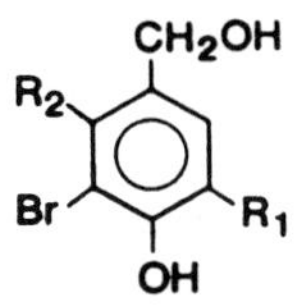 R_1 R_2 Br H OH Br Br Br	*Calothrix brevissima* (blue-green)	(1)
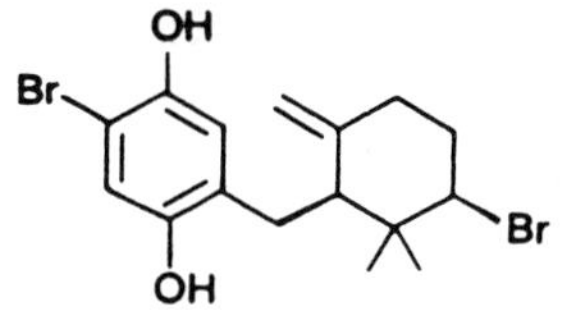 cyclocymopol	*Cymopholia barbata* (green)	(2)
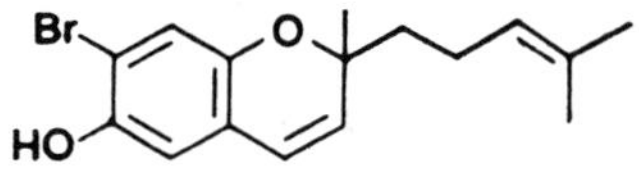 cymopochromenol	*C. barbata* (green)	(2)
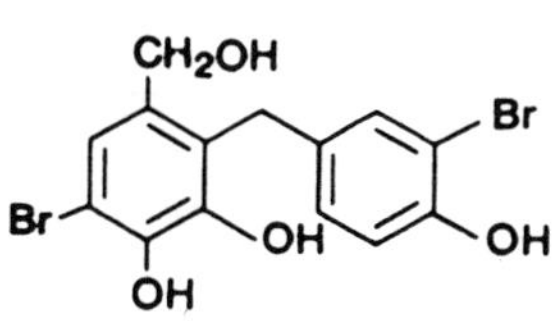 avrainvilleol	*Avrainillea longicaulis* (green)	(3)
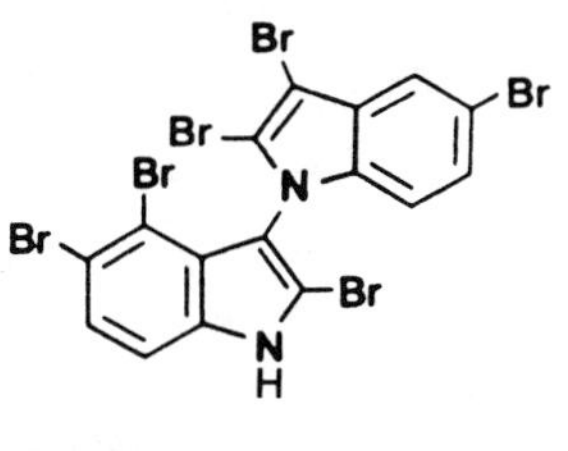	*Rivularia firma* (blue-green)	(4)

(continued next page)

Structure	Source	Ref.
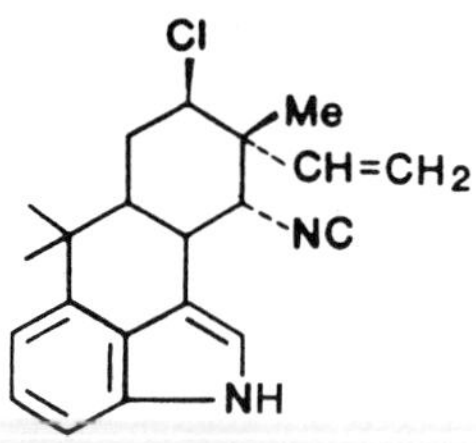 hapalindole A	*Hapalosiphon fontinalis* (blue-green)	(5)
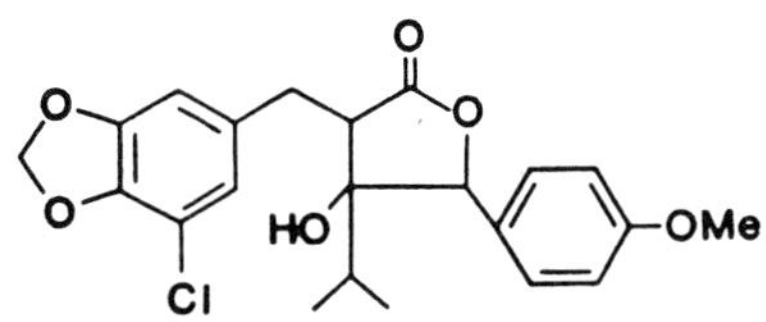 cyanobacterin	*Scytonema hofmanni* (freshwater blue-green)	(6)
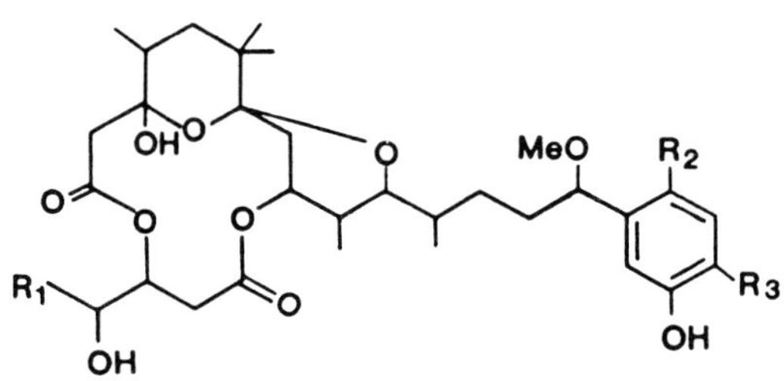 R_1 R_2 R_3 H Br H bromooscillatoxin A H Br Br dibromooscillatoxin A Me Br Br bromoaplysiatoxin	*Oscillatoria nigroviridis* (blue-green)	(7)
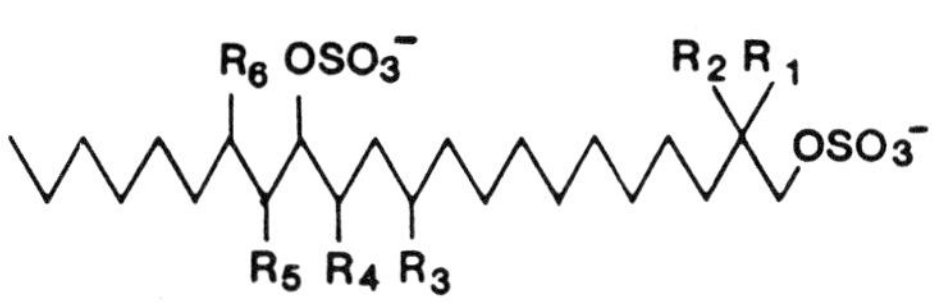 R_1 R_2 R_3 R_4 R_5 R_6 H H H Cl H H H H Cl H Cl H Cl Cl Cl Cl Cl H Cl Cl Cl Cl Cl Cl	*Ochromonas danica* (phytoflagellate)	(8)

(*continued next page*)

Structure	Source	Ref.
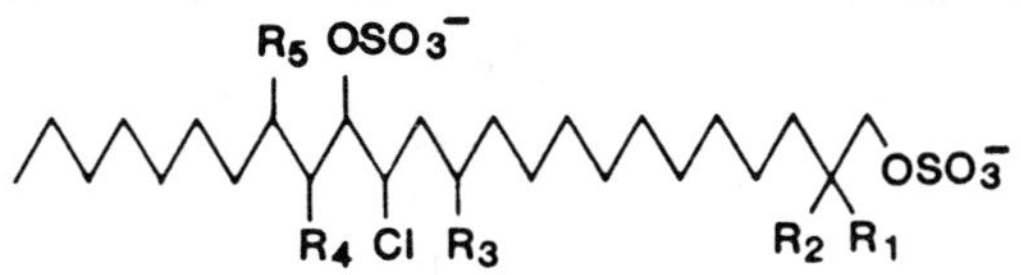	*O. danica*	(8)

R_1	R_2	R_3	R_4	R_5
H	H	H	H	H
Cl	H	Cl	Cl	Cl
Cl	Cl	Cl	Cl	Cl

References:

1. Pederson & Da Silva (1973).
2. Moore (1979).
3. Sun *et al.* (1983).
4. Norton & Wells (1982).
5. Moore *et al.* (1984).
6. Mason *et al.* (1982).
7. Mynderse & Moore (1978).
8. Mercer & Davis (1979).

4. **Higher plants** (Fig. 2.5)

The origin of chlorometabolites in higher plants has been questioned. Are the isolated metabolites really naturally occurring or merely artifacts arising during sample preparation? For example, plant metabolites that contain an epoxide can form minor quantities of chlorohydrins during workup (Hylin *et al.* 1969). However, with properly controlled analytical procedure, it has been confidently demonstrated that the chlorohydrin of withanolide D (Nittala *et al.* 1981) and the chlorinated benzoxazine of young corn roots (Le-Van & Wratten 1984) do indeed occur naturally. In addition, chlorine has been found bound to aromatic nuclei and on aliphatic side chains not adjacent to hydroxyl groups. The presence of chlorine in such structures cannot be explained as artifacts of sample preparation.

At least one source of bromometabolites in a higher plant is known, the brominated stearic acids found in the oil seeds of *Eremostachys molucelloides* (Gusakova & Umarov 1977).

Figure 2.5

Bromo– and chlorometabolities of plants

Structure	Source	Ref.
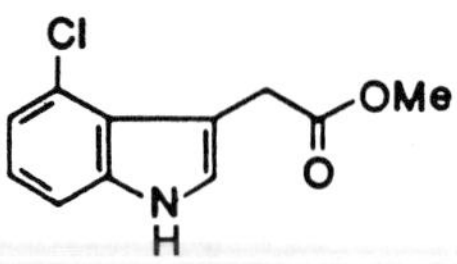	*Vicieae* sp. (immature pea seeds)	(1)
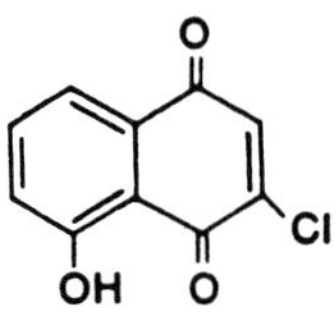	*Plumbago zeylanica* (roots)	(2)
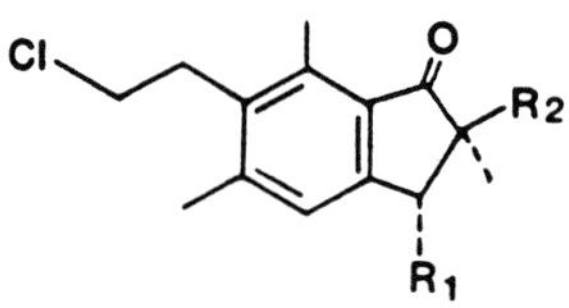 R_1 R_2 OH Me H Me H H	*Pteridium aquilinium* (ferns)	(3)
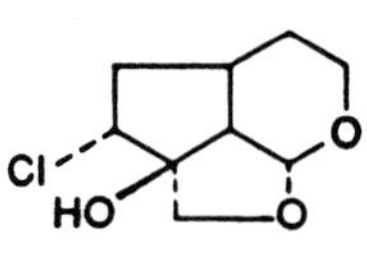 cistachlorin	*Cistanche salsa*	(4)
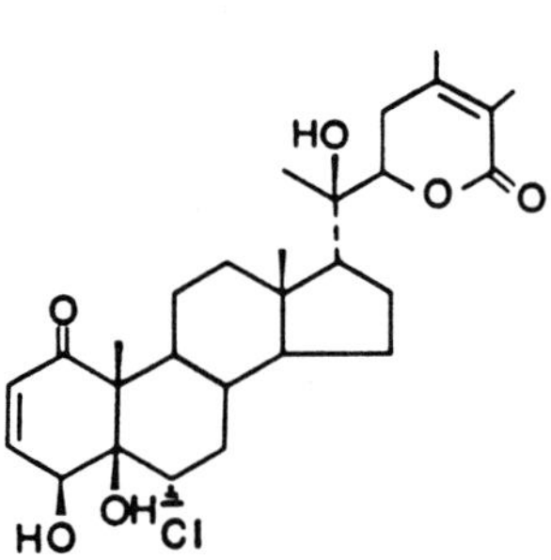	*Withania somnifera*	(5)

Fig. 2.5 – (*continued next page*)

Structure	Source	Ref.
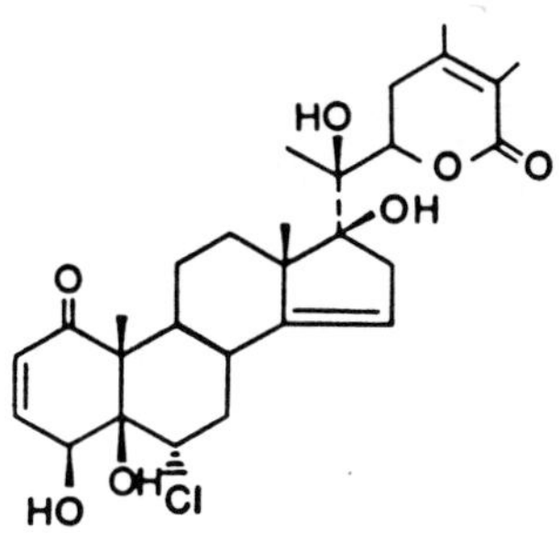	*Physalis peruviana* (leaf)	(6)
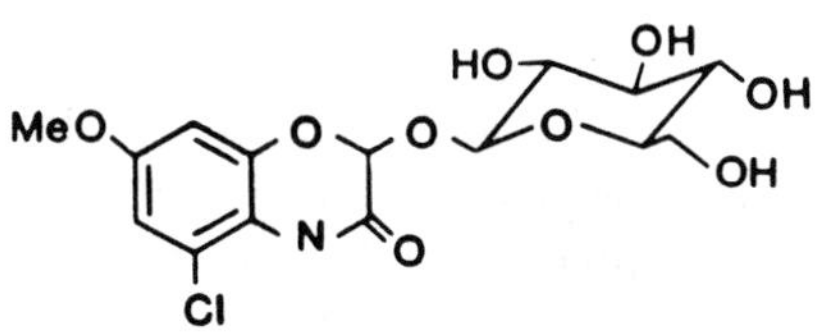	*Zea mays* (young corn roots)	(7)
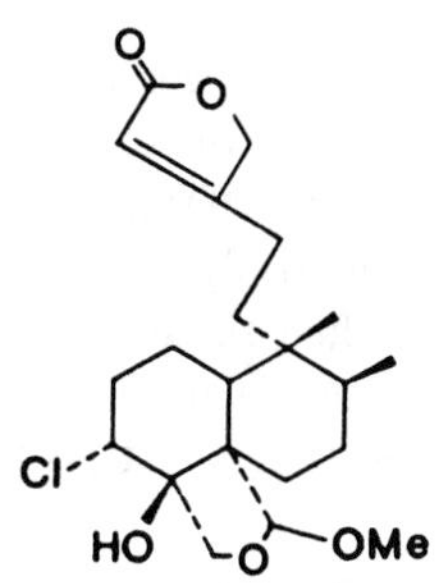	*Gutierrezia dracuncuboides* (herb)	(2)
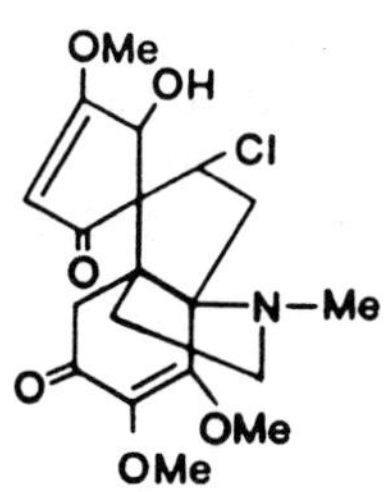	*Sinomenium acutum*	(2)
$CH_3(CH_2)_4\underset{R_1}{C}H\underset{R_2}{C}HCH_2\underset{Br}{C}H\underset{Br}{C}H(CH_2)_7CO_2H$ R_1 R_2 H H Br Br	*Eremostachys molucelloides* (seed oil)	(8)

Fig. 2.5 – (*continued next page*)

References:

1. Engvild *et al.* (1981).
2. Suida & De Bernardis (1973).
3. Tanaka *et al.* (1982).
4. Kobayashi *et al.* (1984).
5. Nittala *et al.* (1981).
6. Ali *et al.* (1984).
7. Le-Van & Wratten (1984).
8. Gusakova & Umarov (1977).

5. Animals (Fig. 2.6)

Marine animals are rich in both bromine and chlorine content (Tinsley & Lowry 1980, Hager 1982), so it is not surprising that an abundance of bromo- and chlorometabolites has been found in the sponges, jellyfish, worms, and sea hares, some of which are shown in Fig. 2.7. The relationship between algal biosynthesis of certain of these compounds and their subsequent appearance in sea hares after algal ingestion will be discussed in Chapter 5.

In this same sense, terrestial animals can accumulate halogenated compounds synthesized by other species – primarily man in his laboratories. The richness of man-made halogen compounds in our environment (e.g., DDT, PCBs, and EDB) provides a ready reservoir for such dietary intake. In one study, when rats were fed brominated stearates (C_{18}) and the bromine level in their tissue lipids measured, not only were there high levels of lipid-bound bromine in their heart and liver tissues, but also there was evidence for shorter chain (C_{16} and C_{14}) lipid-bound bromine (Jones *et al.* 1983a, b).

To date, there has not been any report of organic bromo- or chlorometabolites directly produced by a terrestrial animal.

At the present rate of discovery of new bromo- and chlorometabolites, it is safe to predict that many more compounds remain to be found and identified. Fortunately, access to gas chromatography-mass spectrometry, coupled with the diagnostic mass abundances for bromine (^{79}Br:^{81}Br, 1:1) and chlorine (^{35}Cl: ^{37}Cl, 3:1), now permits their ready detection.

Figure 2.6

Bromo– and chlorometabolites of animals

Structure	Source	Ref.
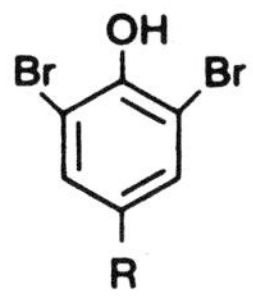		
R		
H	*Balanoglossus biminiensis* (marine worm)	(1)
Br	*Phoronopsis viridis* (marine tube worm)	(2)
CH_2CONH_2	*Verongia archeri* (sponge)	(1)
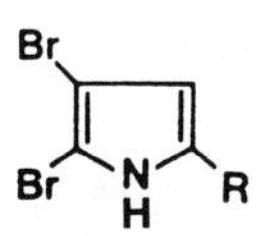		
R		
CN CO_2H $CONH_2$	*Agelas oroides* (sponge)	(1)
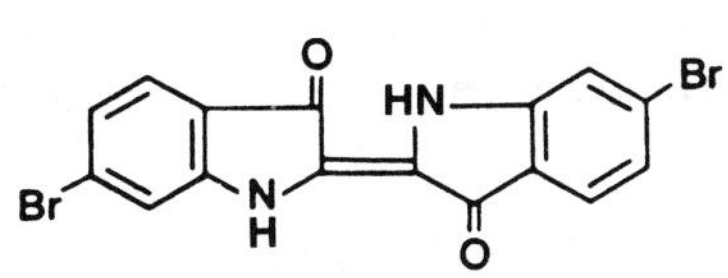		
Tyrian purple dye	*Murex brandaris* (mollusk)	(1)
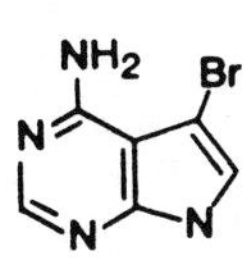	*Echinodictyum* sp. (sponge)	(3)

Fig. 2.6 – *(continued next page)*

Structure	Source	Ref.
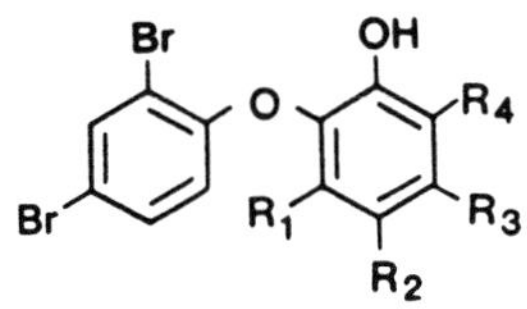		
R_1 R_2 R_3 R_4		
H Br H Br	*Dysidea chlorea*	(4)
Br H Br H	(sponge)	
H Br Br Br	*Dysidea herbacea*	
Br Br Br H	(sponge)	
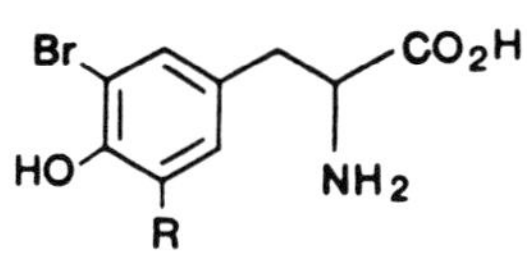		
R		
H	*Despongia* spp.	(1)
Br	(sponge)	
Cl	*Buccinum undatum*	(5)
	(mollusk)	
$CH_3(CH_2)_n\underset{R_1}{C}H\underset{R_2}{C}H(CH_2)_7CO_2H$ n = 5,7		
R_1 R_2		
OH Cl	(jellyfish)	(6)
Cl OH		
$(CH_3)_2CH(CH_2)_{14}CH{=}CH(CH_2)_2\underset{Br}{C}{=}CH(CH_2)_3CO_2H$		
	Petrosia ficiformis	(7)
	(sponge)	
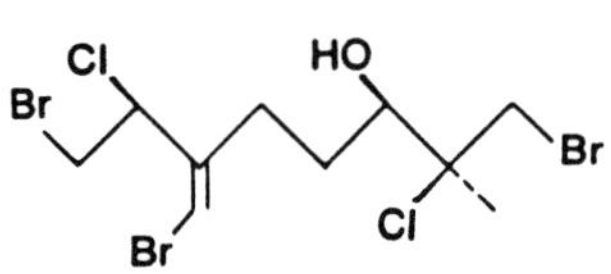		
kurodainol	*Aplysia kurodai*	(8)
	(sea hare)	

Fig. 2.6 – (*continued next page*)

Structure	Source	Ref.
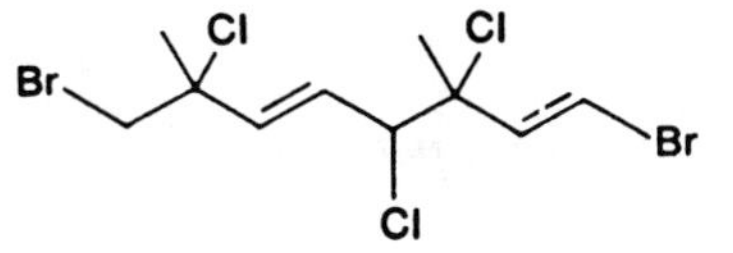	*Aplysia californica* (sea hare)	(9)
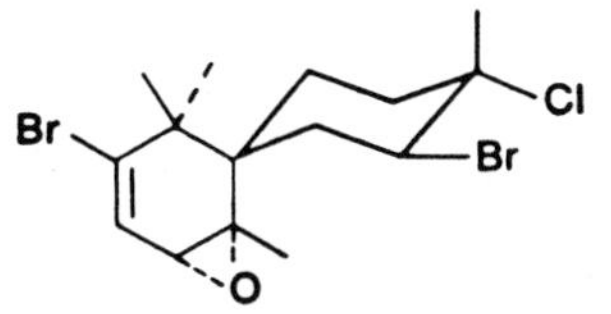	*A. californica*	(9)
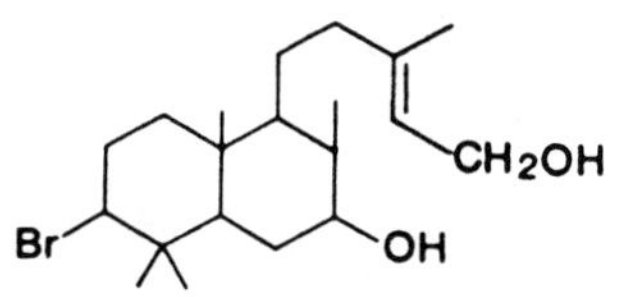 aplysin-20	*Aplysia kurodai*	(1)
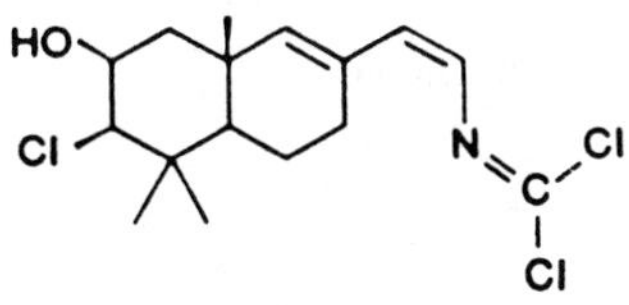	*Pseudaxinysea pitys* (sponge)	(5)
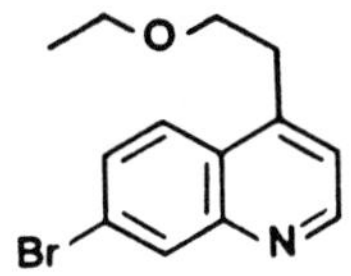	*Flustra foliacea* (marine bryozoan)	(10)
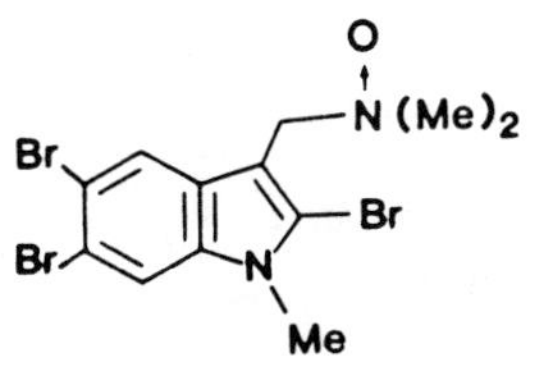	*Zoobotryon verticillatum* (Marine bryozoan)	(11)
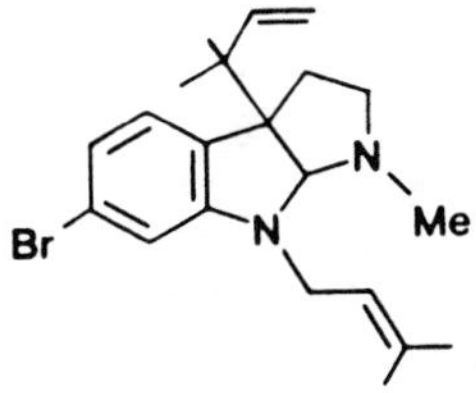	*Flustra folicea*	(5)

Fig. 2.6 – (*continued next page*)

Structure	Source	Ref.
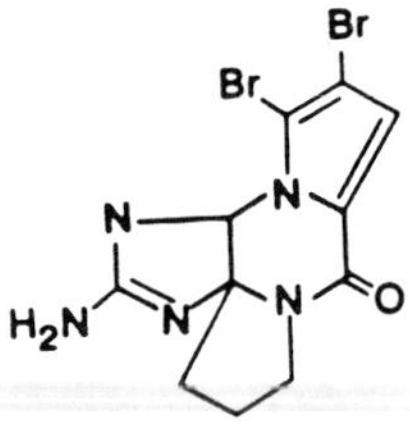	*Phakellia flabellata* (sponge)	(1)
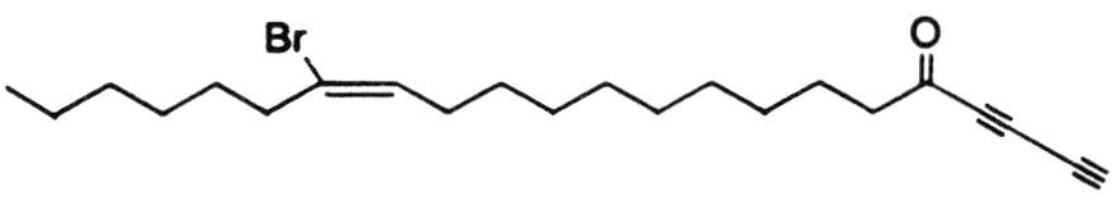	*Reniera fulva* (sponge)	(12)
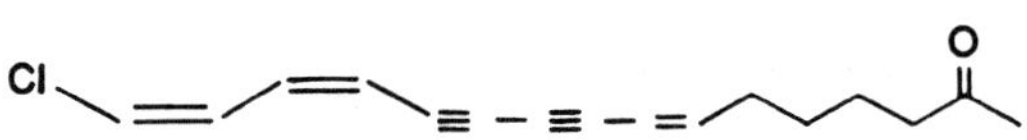	*Diaulula sandiegensis* (nudibranch)	(13)
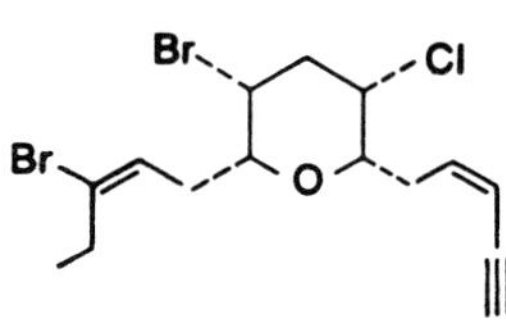	*Aplysia dactylomela* (sea hare)	(12)
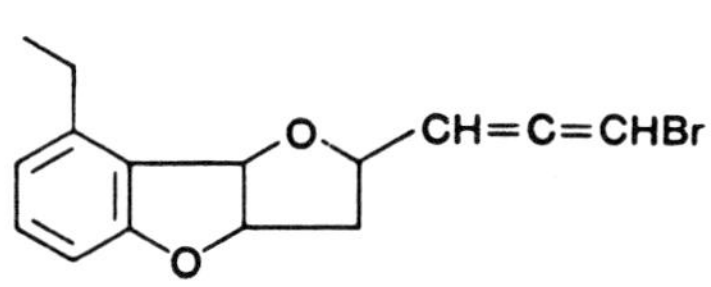 panacene	*Aplysia brailiana* (sea hare)	(14)

References:

1. Suida & DeBernardis (1973).
2. Sheikh & Djerassi (1975).
3. Kazlauskas *et al.* (1983).
4. Carte & Faulkner (1981).

Fig. 2.6 – (*continued next page*)

5. Fenical (1982).
6. White & Hager (1977).
7. Wijekoon *et al.* (1984).
8. Katayama *et al.* (1982).
9. Ireland *et al.* (1976).
10. Wulff *et al.* (1982).
11. Sato & Fenical (1983).
12. Thomson (1978).
13. Walker & Faulkner (1981).
14. Kinnel *et al.* (1977).

IODOMETABOLITES

While iodine is not as abundant as chlorine or bromine in the environment (only $3 \times 10^{-5}\%$ by weight in rocks, and $0.05\,\mu g/ml$ in seawater), iodometabolite formation is still quite prevalent. As shown in Fig. 2.7, many of these compounds have been found in red algae. This is not surprising since iodine concentrations as high as 0.5% of wet weight are known in these algae (Fenical 1975, Whyte & Englar 1976). However, the best-known iodometabolite is the hormone present in the human bloodstream, thyroxine. It is maintained at concentrations around 10^{-9} molar, with very slight variations from this level causing dramatic physiological effects. Thyroxine has also been reported from invertebrates and algae (Hunt 1984).

High iodide intake can result in increased iodometabolite formation. Thyroid glands from dogs fed on high, normal, and low iodide diets were examined. Iodinated lipids were readily detected in the thyroid glands from dogs fed the high iodide diet (Rabinowitz *et al.* 1976).

The relative instability of the carbon–iodine bond may explain why many more iodometabolites have not yet been discovered. Nucleophilic exchange with OH^- ions can readily result in a loss of iodine during sample preparation. The facile rupture of the carbon–iodine bond by thermal energy, for example, in the injector port of a gas chromatograph, is another way by which iodine can be lost from compounds during analysis.

Figure 2.7

Idometabolites and their sources

Structure	Source	Ref.
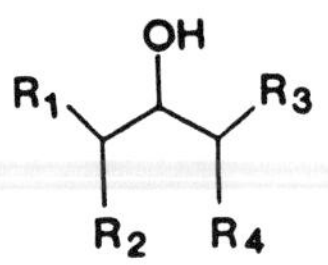 R_1 R_2 R_3 R_4 H Cl H I H Br H I H I H I Cl Br H I Br Br H I Br Br Br I Br Br I I	*Asparagopsis* spp. (red algae)	(1)
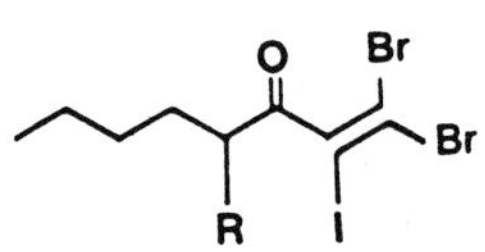 R H Cl	*Delisea fimbriata* (red alga)	(2)
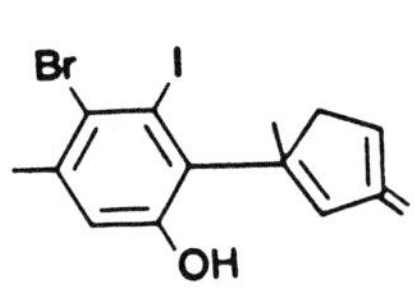	*Laurencia nana* (red alga)	(3)
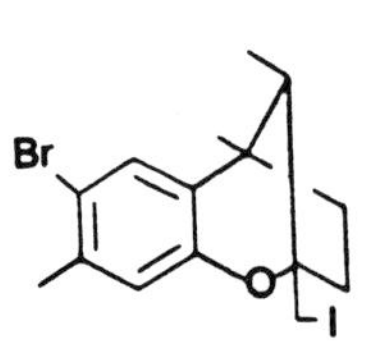	*Laurencia nana* (red alga)	(3)

Fig. 2.7 – (*continued next page*)

Structure	Source	Ref.
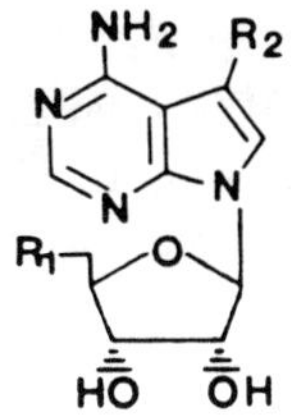 R_1 R_2 H I OH I	*Hypnea valendiae* (red alga)	(4)
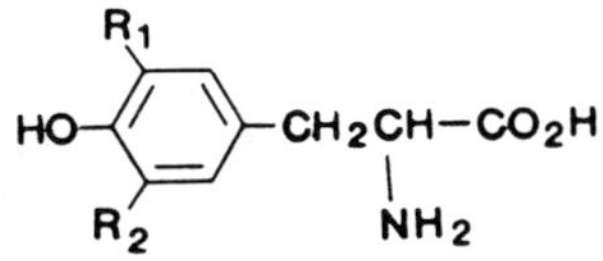 R_1 R_2 H I I I	(numerous, algae, corals, sponges, animals)	(5)
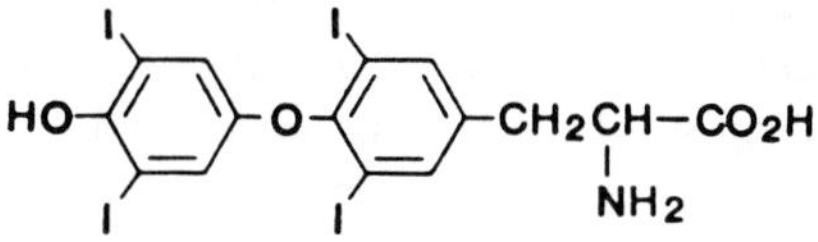 thyroxine	(mammals)	(5)
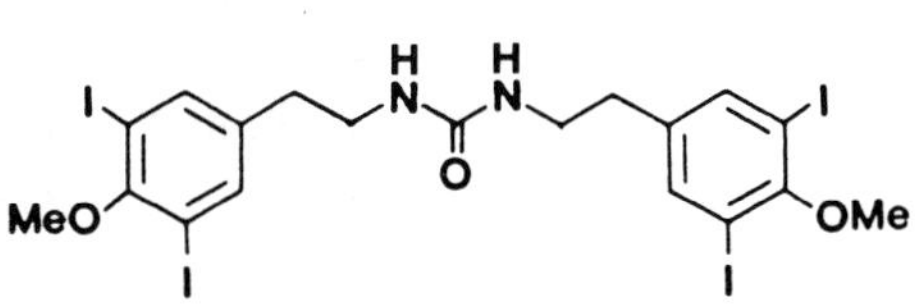	*Didemnum* sp. (tunicate)	(6)

References:

1. Moore (1978).
2. Rose *et al.* (1977).
3. Izac & Sims (1979).
4. Kazlauskas *et al.* (1983).
5. Suida & DeBernardis (1973).
6. Sesin & Ireland (1984).

FLUOROMETABOLITES

Although fluorine is not an abundant halogen in the environment, locally high fluoride concentrations do occur in nature. Plants growing in areas containing either fluoride-rich bedrock or fluorspar (CaF_2) mining waste have been found to contain levels of fluoride ranging between 100 and 10 000 ppm dry weight (Cooke *et al.* 1976, Takala *et al.* 1979). Fluoride constitutes about 10% of the dry weight of the marine sponge *Halichodria moorei*, principally as potassium fluorosilicate (Gregson *et al.* 1979), even though fluoride ion concentration in seawater is less than 2 ppm. Fluoride accumulation appears to be a common property of oral microorganisms, with a two- to six-fold concentration reported from a 1 ppm fluoride solution, which is the concentration in fluorinated drinking water (Gaugler *et al.* 1981). One wonders what happens when these same microorganisms are exposed to fluoride-containing toothpastes containing about 1000 ppm fluoride. The basal concentration of fluoride in urine is normally under 1 ppm; however, elevated levels of inorganic fluoride up to 60 ppm appear in the urine of patients who undergo treatment with the fluorocarbon inhalation anaesthetic, methoxyfluorane, $Cl_2CHCF_2OCH_3$ (Phillips & Rix 1981). These localized high fluoride concentrations, resulting from both natural and artificial exposures, could provide opportunities for fluorometabolite biosynthesis.

The known fluorometabolites are shown in Fig. 2.8. Fluoroacetate was the first fluorometabolite discovered, found over 40 years ago in the poisonous African plant, *Dichapetalum cymosum* (Marais 1944). Since then, fluoroacetate has been detected in other plants growing in Africa, South America, and Australia (Hall 1977). Fluoroacetate can occur in concentrations up to 2000 ppm (air-dried tissue) in these plants. Why and how these plants synthesize this compound are intriguing questions. Two mechanisms have been proposed, but both are still unproven. The first mechanism involves the enhancement by pyridoxal phosphate of nucleophile (fluoride ion) incorporation into an amino acid such as cysteine or serine. Transamination, decarboxylation, and oxidation then take place resulting in the fluorinated acid (Mead & Segal 1972). The other mechanism involves fluorodecarboxylation of malonic acid by a fluoroperoxidase (Vickery *et al.* 1979). This latter postulate will be negated in the discussion in Chapter 4. The recent success in establishing callus tissue cultures of *Acacia georginae* now permits the further study of fluoroacetate synthesis by a laboratory plant model (Bennett *et al.* 1983).

Other fluorinated compounds are also present in fluoroacetate-producing plants. It has been demonstrated that *Glycine max* (soya bean) converts fluoroacetate to fluorocitrate, most likely through the activity of the enzyme, citrate (si)-synthase (EC 4.1.3.7), which couples fluoroacetyl CoA and oxaloacetate (Brandange *et al.* 1982). Four fluorinated fatty acids were found in the seed oil of *Dichapetalum toxicarium*. This oil contained up to 3% by weight of ω-fluorooleic acid with a lesser amount of ω-fluoropalmitic acid. Evidence further suggested that the oil also contained ω-fluorocapric acid and ω-fluoromyristic acid, but these compounds were not isolated (Ward *et al.* 1964).

The microorganisms *Streptomyces calvus* was shown in 1957 to produce an antibiotic substance, nucleocidin, related to the nucleic acid adenosine.

However, it was not until 1969 that it was finally realized the nucleocidin actually contained fluorine (Jenkins *et al.* 1976). There have been no mechanisms proposed to explain the biosynthesis of this unique fluorocarbohydrate. Even the source of the fluoride for the synthesis of nucleocidin is not clear since fluoride was not specifically added to the fermentation medium (Thomas *et al.* 1959). Trace contamination of fluoride ion in the medium salts of the water appears to be its source.

The 12-year gap between the discovery of nucleocidin and the detection of fluorine in the molecule illustrates the difficulty in detecting fluorometabolites. Without the modern-day tools of analytical chemistry (e.g., ^{19}F-NMR, mass spectrometry, ion chromatography, F^--selective electrodes), the biochemist or natural product chemist would be at a loss to readily distinguish between a C–H bond and a C–F bond. These tools are also now being used to screen natural sources for new organisms that produce known fluorinated compounds (Vartiainen & Kauranen 1984). Hopefully, the multidisciplinary interaction between chemists and biologists will lead to detection of other fluorometabolites.

Figure 2.8

Fluorometabolites

Structure	Source	Ref.
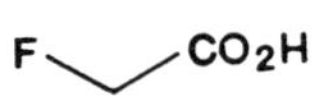 fluoroacetate	(numerous higher plants)	(1)
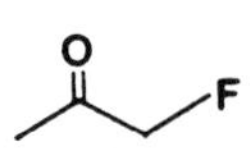 fluoroacetone	*Acacia georginae* (higher plant)	(2)
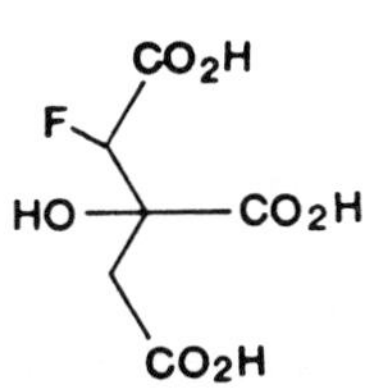 fluorocitrate	(several higher plants)	(2)

(continued next page)

Structure	Source	Ref.
$F-(CH_2)_nCO_2H$	*Dichapetalum toxicarium* (higher plant)	(1)
n		
9		
13		
15		
$F-(CH_2)_8CH=CH(CH_2)_7CO_2H$		
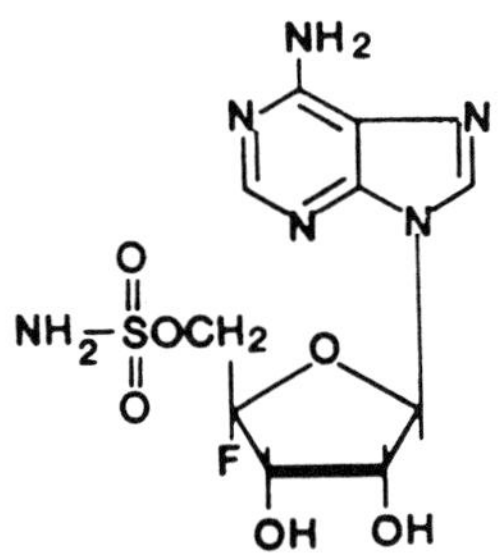 nucleocidin	*Streptomyces calvus* (bacterium)	(3)

References:

1. Suida & DeBernardis (1973).
2. Peters & Shorthouse (1972).
3. Jenkins *et al.* (1976).

CONCLUSION

The attitude concerning the abundance of halometabolites has, indeed changed over time. We have come a long way in 25 years since the conclusion reached by Petty (1961) that halogens are a relatively uncommon structural feature among metabolites. It is certain that the number of halometabolites reported from all sectors of the animal and plant kingdoms will continue to increase. Table 2.4 indicates the situation at present with respect to the occurrence of various types of halometabolites in nature. It is likely that all the empty slots in the table will eventually be filled.

Succeeding chapters will be devoted to considering two major areas that require further research: 1) what are the *in vivo* mechanisms for halogenation in nature, and 2) what biological role is associated with these diverse halometabolites. The first question will be considered in Chapters 3 and 4; the second, in Chapters 5 and 6.

Table 2.4 – Specific halometabolite classes reported to be produced by various biological groups, as of 1985.

	Halometabolite discovered			
Group	F-Metabolite	Cl-Metabolite	Br-Metabolite	I-Metabolite
Bacteria	yes	yes	yes	no
Fungi	no	yes	no	no
Algae	no	yes	yes	yes
Higher plants	yes	yes	yes	no
Higher animals	no	yes	yes	yes
Humans	no	no	no	yes

3

The halogenating enzymes (haloperoxidases): sources and reactions

INTRODUCTION

Although an impressive number of halometabolites have been isolated and identified, there has been only one type of halogenating enzyme isolated to date: haloperoxidase. We will show in this chapter that haloperoxidases are widely distributed throughout nature and are capable of carrying out a multitude of reactions according to the following to the following general equation:

$$\text{substrate} + H_2O_2 + X^- + H^+ \xrightarrow{\text{enzyme}} \text{halogenated product} + 2H_2O$$

The haloperoxidases are grouped according to the range of halide ions, X^-, that they can utilize:

Range of halide ions	*Group name*
Cl^-, Br^-, I^-	chloroperoxidase
Br^-, I^-	bromoperoxidase
I^-	iodoperoxidase

HALOPEROXIDASES IN NATURE

Haloperoxidases are peroxidases that catalyze halogenation. Although peroxidases were known before 1900, the discovery of the halogenating peroxidases occurred much later. By the early 1950s, only three haloperoxidases had been discovered: horseradish peroxidase from the horseradish root (EC 1.11.1.7, HRPO), lactoperoxidase from milk, saliva, and tears (EC 1.11.1.7, LPO), and myeloperoxidase from white blood cells (EC 1.11.1.7, MPO), also referred to as verdoperoxidase in the early literature owing to its dark green color (Saunders 1957). By the 1960s, an additional three haloperoxidases were identified: thyroid peroxidase from the thyroid gland (TPO), chloroperoxidase from the fungus *Caldariomyces fumago* (EC 1.11.1.10, CPO), and eosinophil peroxidase (EPO) from white blood cells (Alexander 1959, Shaw & Hager 1959, Archer *et al.* 1965). However, it has been in the past ten years that the broad occurrence of haloperoxidases became apparent. During the mid-1970s the Scripps Institution of Oceanography sponsored an expedition to collect and survey marine organisms for the presence of halometabolites. From over 1000 organisms examined, 50 marine red and green algae were discovered to produce bromoperoxidases (Theiler *et al.* 1978a, Hewson & Hager 1980). During the early 1980s, Cetus Corporation sponsored a tour to Death Valley to collect microbial samples. From these samples, over 80 fungal species, all belonging to the family of dematiaceous hyphomycetes, were discovered to produce chloroperoxidases (Hunter *et al.* 1985a, b, Geigert *et al.* 1983a). Also added to the growing list of new haloperoxidases are the following:

1. bromoperoxidases in marine worms (Ashern *et al.* 1980)
2. bromoperoxidase in sea urchin eggs (Deits *et al.* 1984)
3. the first haloperoxidases (both bromoperoxidases) discovered in bacteria (van Pee & Lingens 1984, van Pee & Lingens 1985)
4. iodoperoxidases in marine brown algae (Vilter *et al.* 1983)
5. iodoperoxidases in American wormseed fruit (Johnson & Croteau 1984).

A summary of the known sources for haloperoxidases, grouped according to their ability to oxidize halide ions, is presented in Tables 3.1 through 3.3. In addition to the iodoperoxidases listed, an iodoperoxidase-like enzyme has been reported in the marine bacterium, *Pseudomonas iodooxidans.* However, this enzyme could only oxidize iodide in the presence of a polysaccharide such as starch, cellulose, or glycogen (Gozlan & Margalith 1974). The role of the polysaccharide is not understood. Also, there have been two reports claiming discovery of chloroperoxidases in marine organisms: *Cystoclonium purpureum* (Pedersen 1976) and *Iotrochota birotulata* (Baden & Corbett 1979). However, the red alga *C. purpureum* could only incorporate chloride in the presence of bromide and only in the presence of substrates that lost formaldehyde during the halogenation reaction. Chlorination by the sponge *I. birotulata* was measured only by the monochlorodimedon assay, which is known not to be specific for chloride ion uptake. In addition, both of these peroxidases exhibited atypical characteristics of a chloroperoxidase: *C. purpureum* could not oxidize iodide, while *I. birotulata* could not oxidize bromide. Therefore, the true nature of these peroxidases must await further studies.

Table 3.1 – Sources of iodoperoxidases.

1. Mammals and birds	
thyroid gland	TPO
2. Plants	
horseradish root	HRPO
turnip root	
Chenopodium ambrosioides (wormseed) fruit	
3. Brown algae (Phaeophyta)	
Desmarestia aculeata	
Fucus serratus	
Fucus spiralis	
Fucus vesticulosus	
Halopteris scoparia	
Laminaria digitata	
Laminaria hyperborea	
Laminaria saccharina	
Pelvetia canalicutata	

Table 3.2 – Sources of bromoperoxidases.

1. Mammals		
secretions of exocrene glands:		
milk, saliva, tears		LPO
2. Marine animals		
Thelepus setosus (worm)		
Ptychodera flava laysanica (worm)		
Strongylocentrotus purpuratus (sea urchin) egg		OVOPEROXIDASE
3. Marine algae		BPO
a. Rhodophyta (red)		
Amphiroa hancockii	*Galaxaura squalida*	
Asparagopsis taxiformis	*Gracilaria* sp.	
Bonnemaisonia hamifera	*Haloplegma duperreyi*	
Bryothamnion triquetrum	*Laurencia chondroides*	
Chondria atropurpurea	*Laurencia intricata*	
Chondria cnicophylla	*Laurencia lata*	
Coelothrix irregularis	*Laurencia obtusa*	
Dasya pedicellata	*Laurencia papillosa*	
Dasya sp.	*Laurencia poitei*	
Euchewna isiforme	*Liagora* sp.	
Galaxaura comans	*Ochtodes secundiramea*	
b. Chlorophyta (green)		
Bryopsis pennata	*Halimeda opuntia*	
Caulerpa racemosa	*Halimeda tuna*	
Chaetomorpha linum	*Penicillus capitatus*	
Cladophorales sp.	*Penicillus dumentosus*	
Caldophoropsis membranacea	*Penicillus lamourouxii*	
Codium isthmocladum	*Rhipocephalus phoenix*	
Codium repens	*Udotea flabellum*	
Dictyosphaeria cavernosa	*Udotea wilsoni*	
Halimeda discoidae	*Ulva lactuca*	
Halimeda incrassata	*Valonia macrophysa*	
Halimeda goreauii		
c. Phaeophyta (brown)		
Ascophyllum nodosum		
Dictyota divaricata		
Lobophora variegata		
4. Bacteria		
Pseudomonas aureofaciens		
Streptomyces phaeochromogenes		

Table 3.3 – Sources of chloroperoxidases.

Source	Enzyme
1. Mammals	
white blood cells	
subset neutrophils	MPO
subset eosinophils	EPO
2. Fungi	
Dematiaceous hyphomycetes	
Alternaria alternata	
Bispora betulina	
Botrytis cinerea	
Caldariomyces fumago	CPO
Cladosporium cladosporioides	
Cladosporium resinae	
Cladosporium sphaerospermum	
Curvularia cymbopogonis	
Curvularia inaequalis	
Curvularia lunata	
Curvularia tuberculata	
Curvularia verruciformis	
Dendryphiella salina	
Dendryphion nanum	
Dichotomphthora portulacae	
Drechslera halodes	
Embellisia allii	
Epiccocum nigrum	
Gliocephalotrichum sp.	
Helicoma isiola	
Humicola fuscoatra	
Stachybotrys sp.	
Stemphyllium macrosporidium	
Ulocladium chartarum	
Wallemia sebi	
Wardomyces anomalus	

Although ecological niches have been demonstrated for each haloperoxidase group (e.g., dematiaceous hyphomycetes for chloroperoxidases, red algae for bromoperoxidases, and brown algae for iodoperoxidases), the abundance and variety of halometabolites listed in Chapter 2 indicate that other niches remain to be discovered. Where are the chlorinating enzymes that must exist in the *Moniliaceae* (e.g., species of *Aspergillus* and *Penicillium*, mentioned in Table 2.2) to account for the chlorometabolites produced by over 100 members of this fungal family? Where are the algal chlorinating enzymes that produce the many chlorometabolites listed in Figs. 2.3 and 2.4? Where are the chlorinating enzymes involved in the bacterial biosynthesis of the commercially important antibiotics (e.g., chlortetracycline and chloramphenicol)?

ISOLATION AND PURIFICATION

To effectively study the nature and reactivity of an enzyme, it must be isolated from its natural environment and purified. Various isolation and purification methods have been applied to haloperoxidases as illustrated by Tables 3.4 through 3.11. The natural environment in which one finds haloperoxidases is often complex; for example, fungal mycelia, milk, pus, and blood. Fortunately, an abundance of separation methods are available for enzyme purification. The strategies outlined in Tables 3.4 through 3.11 involve elements of the following separation methods:

Desired effect	*Separation method used*
Release of enzyme from cell	Homogenization Detergent solubilization Solvent extraction Enzyme digestion
Solid–liquid separation	Filtration Centrifugation Salt precipitation Solvent precipitation
Concentration	Evaporation Ultrafiltration Dialysis
Chromatographic fractionation	Adsorption Gel filtration Ion-exchange Reverse-phase Affinity

Table 3.4 – Purification and crystallization of chloroperoxidase (CPO) from the fungus, *Caldariomyces fumago*.

Purification from fungus

1. Filter the mold mycelium from 30 liters of fermentation broth.*
2. Concentrate the filtrate under a vacuum.
3. Dialyze the concentrated material against deionized water.
4. Lyophilize.
5. Dissolve the freeze-dried powder in distilled water and add ethanol to bring the solution to 45% (v/v) saturation at −10°C.
6. Centrifuge and save supernatant.
7. Add additional ethanol to bring the solution to 65% saturation.
8. Decant supernatant and save precipitate.
9. Dissolve the gelatinous precipitate in 0.1 M phosphate buffer (pH 6.0).
10. Apply the chloroperoxidase solution to a column containing DEAE-Sephadex A-50.
11. Wash the ion exchange column with 0.1 M phosphate buffer (pH 6.0).
12. Elute chloroperoxidase with a linear ionic strength gradient of 0.1 to 0.2 M phosphate buffer (pH 6.0).
13. Dialyse the chloroperoxidase fractions against 0.01 M phosphate buffer (pH 5.0).
14. Apply the chloroperoxidase fractions to a column containing calcium phosphate gel-cellulose.
15. Elute chloroperoxidase with a linear gradient of 0.01 M to 0.2 M phosphate buffer (pH 5.0).

    (A yield of about 10 mg of CPO was obtained. The enzyme purity was about 75%.)

Crystallization

1. Add ammonium sulfate to 70% (wt/v) saturation.
2. Centrifuge and save precipitate.
3. Dissolve precipitate in a minimal amount of 0.01 M phosphate buffer (pH 6.0).
4. Dialyze against 60% saturated ammonium sulfate.
5. Add ammonium sulfate to 70% saturation at 4°C and wait.

    (Crystals of CPO appeared within 1 day.)

* Fermentation procedures have been described by Pickard (1981).

Reference: Morris & Hager (1966b).

Table 3.5 – Purification of chloroperoxides from dematiaceous hyphomycetes.

1. Sonicate the fermentation broth.*
2. Filter the growth material from the solution.
3. Add ammonium sulfate to the filtrate to 40% (wt/v) saturation.
4. Centrifuge and save supernatant.
5. Add additional ammonium sulfate to the supernatant to bring it up to 55% saturation.
6. Centrifuge and save precipitate.
7. Dissolve precipitate in 0.1 M potassium phosphate (pH 6.0).
8. Load the solution on a DEAE-Sepharose CL-6B column.
9. Elute the chloroperoxidase from the ion exchange column with a linear gradient of 0.05 M to 0.5 M potassium phosphate (pH 6.0).
10. Load the enzyme fraction on a Sephacryl S-300 column.
11. Elute the chloroperoxidase from the gel filtration column with potassium phosphate pH 6.0.

* Fermentation procedures have been described by Hunter *et al.* (1985b).

Reference: J. Geigert *et al.* (1985).

Table 3.6 – Purification and crystallization of myeloperoxidase (MPO) from canine pus.

Purification from canine pus

1. Solubilize the peroxidase by extracting 1 liter of canine pus with 1% cetyltrimethylammonium bromide in 0.1 M K_2HPO_4.
2. Centrifuge and save the supernatant.
3. Precipitate the soluble myeloperoxidase in the supernatant by addition of ammonium sulfate to 55% (wt/v) saturation.
4. Centrifuge and save the precipitate.
5. Dissolve the pellet in 0.05 M K_2HPO_4 and dialyze.
6. Load the myeloperoxidase solution on a column containing Amberlite CG-50.
7. Wash the ion exchange column with 0.1 M K_2HPO_4.
8. Elute myeloperoxidase with 0.2 M K_2HPO_4.
9. Load the myeloperoxidase fraction on a column containing hydroxyapatite.
10. Wash the adsorption column with 0.1 M K_2HPO_4.
11. Elute myeloperoxidase with a linear gradient of 0.1 to 0.5 M K_2HPO_4.

   (A yield of about 500 mg of MPO was obtained; approximately 60% overall recovery. The enzyme purity was greater than 90%.)

Crystallization

1. Dialyze against 0.2 M K_2HPO_4.
2. Slowly add ethyl alcohol to 30% (v/v) at 4°C and wait.

   (Crystals of MPO appeared within 3 days.)

Reference: Schultz (1980).

Table 3.7 – Purification of myeloperoxidase (MPO) from a continuous cell line.

1. Transplant subcutaneously human promyelocytic leukemia HL-60 cells into the backs of male athymic nude mice.
2. Harvest the solid tumors about 2 months later.
3. Chop up the tumors (~18 g) with scissors and then homogenize in 60 ml of 6.7 mM sodium phosphate buffer (pH 6.6), containing 3 mM NaCl, 1 mM $MgCl_2$, 0.5 mM phenylmethanesulfonyl fluoride and 0.3% Triton X-100.
4. Filter the homogenate on gauze and then centrifuge the filtrate. Save the pellet.
5. Wash the pellet with 60 ml of 6.7 mM sodium phosphate buffer (pH 6.6) containing 3 mM NaCl, 1 mM $MgCl_2$, and 0.5 mM phenylmethanesulfonyl fluoride, and then centrifuge. Save the pellet.
6. First homogenize and then extract the washed pellet with 80 ml of 1% cetyltrimethylammonium bromide in 0.1 M potassium phosphate (pH 7.8). Repeat and combine extracts.
7. Load the extract on a CM-Sepharose CL-6B column.
8. Wash the column with 0.1 M potassium phosphate (pH 7.8) containing 0.02% cetyltrimethylammonium bromide.
9. Elute myeloperoxidase with a linear gradient of 0.1 M to 0.5 M potassium phosphate (pH 7.8) containing 0.02% cetyltrimethylammonium bromide.
10. Precipitate the enzyme by bringing the solution to 70% (wt/v) saturation of ammonium sulfate.
11. Centrifuge to obtain a pellet.
12. Resuspend the pellet in 0.02 M potassium phosphate pH 7.8.

    (A yield of about 40 mg of MPO was obtained; approximately 70% overall recovery.)

Reference: Yamada *et al.* (1981).

Table 3.8 – Purification and separation of myeloperoxidase (MPO) and eosinophil peroxidase (EPO) from human blood.

Purification from human blood

1. Buffy coats obtained by centrifuging 90 liters of outdated human blood.
2. Wash the buffy coats twice in a hypo-osmotic solution to lyse contaminating erythrocytes.
3. Centrifuge to sediment the white cells.
4. Resuspend the packed white cells in 0.1 M potassium phosphate buffer (pH 7.3) and homogenize; centrifuge and save pellet and supernatant. Repeat two more times.
5. Resuspend the pellet in 1.5 liters of 0.5 M $CaCl_2$ and homogenize; centrifuge and save supernatant. Repeat once more and combine supernatants.
6. Add 0.1 volume of 1 M sodium acetate buffer (pH 5.6) containing 0.5% (w/v) cetyltrimethylammonium bromide, 0.5 M $CaCl_2$, 0.05 M $MgCl_2$.and 0.05 M $MgCl_2$.
7. Mix the haloperoxidase preparation with concanavalin A-Sepharose and load into a column.
8. Wash the affinity column with 0.1 M sodium acetate buffer (pH 5.6) containing 0.05% cetyltrimethylammonium bromide, 0.05 M $CaCl_2$, 5 mM $MnCl_2$, 5 mM $MgCl_2$, and 25% (w/v) glycerol (referred to as Buffer A).
9. Elute the haloperoxidases with Buffer A which also contains 0.2 M 1-O-methyl-α-D-glycopyranoside.
10. Concentrate the haloperoxidase fractions to about 100 ml by ultrafiltration.
11. Load the haloperoxidase solution on a column containing phenyl-Sepharose.
12. Wash the adsorption column with Buffer A.
13. Elute the haloperoxidases with Buffer A containing 75% (w/v) (rather than 25%) glycerol.
14. Concentrate the haloperoxidase fractions to about 4 ml by ultrafiltration.

   (A yield of about 200 mg of MPO + EPO was obtained*; approximately 35% overall yield.)

Separation of MPO from EPO

1. Load the purified haloperoxidase solution on a gel filtration column containing Ultrogel AcA 44.
2. Elute the haloperoxidases with Buffer A.

   (MPO elutes before EPO; approximately a 10:1 weight ratio of haloperoxidase yield.)

* Assay of MPO and EPO in mixtures can be accomplished with the use of 3-amino-1,2,4-triazole (AMT) in the assay reaction. AMT completely inhibits the activity of EPO, but only slightly affects the activity of MPO (Cramer *et al.* 1984).

Reference: Olsen & Little (1983).

Table 3.9 – Purification and separation of lactoperoxidase (LPO) and non-heme lactoperoxidase from bovine milk.

1. Remove casein from fresh raw skim bovine milk.
2. Coagulate the milk by adding rennet and stir at room temperature.
3. Centrifuge and save the whey.
4. Add Amberlite CG-50 (NH_4^+ form) to the whey and stir for 3 hours.
5. Let the resin settle and then transfer the resin to a Buchner funnel.
6. Wash the resin first with water and then with 50 mM sodium acetate.
7. Transfer the resin to a column.
8. Elute each lactoperoxidase with the following solutions:
   (a) LPO – 0.5 M sodium acetate
   (b) non-heme LPO – 1.0 M sodium acetate.
9. Precipitate the separated lactoperoxidases with 50% (wt/v) ammonium sulfate.
10. Centrifuge and save the precipitates.
11. Resuspend the precipitates in 100 mM sodium acetate.
12. Load each lactoperoxidase solution on a Sephadex G-25 column.
13. Elute with 100 mM sodium acetate to separate the enzymes from the ammonium sulfate.

    (A yield of about 20 mg LPO and 35 mg of non-heme LPO were obtained per liter of whey; approximately 90% overall recovery.)

Reference: Dumontet & Rousset (1983).

Table 3.10 – Purification of bromoperoxidase from the marine green alga, *Penicillus capitatus.*

1. Remove sand and debris from 2.5 kg of algae.
2. Cut the stalks and capitula into 1-inch segments and homogenize in 0.1 M sodium phosphate buffer (pH 6.8).
3. Filter the homogenate through cheesecloth and centrifuge.
4. Concentrate the supernatant using an Amicon hollow reed concentrator.
5. Add ammonium sulfate to bring the solution to 60% (wt/v) saturation.
6. Centrifuge at 12 000 g for 1 hr and discard pellet.
7. Add additional ammonium sulfate to bring the solution to 60% saturation.
8. Centrifuge and save pellet.
9. Dissolve the pellet in 0.25 M Tris buffer (pH 7.0) and dialyze against 0.1 M Tris buffer (pH 7.0).
10. Load the bromoperoxidase solution on a column containing DEAE-Sephadex A-50.
11. Wash the gel filtration column with 0.3 M Tris buffer (pH 7.0).
12. Elute bromoperoxidase with a linear gradient of 0.1 to 0.5 M Tris buffer (pH 7.0).
13. Concentrate the bromoperoxidase fractions using an Amicon hollow reed concentrator and then dialyze.
14. Repeat steps 10–12.
15. Load the solution on a column containing Sephadex G-200.
16. Elute bromoperoxidase with 0.1 M potassium phosphate buffer (pH 7.0).

    (A yield of about 5 mg of the bromoperoxidase was obtained; approximately 30% overall recovery.)

Reference: Manthey *et al.* (1984).

Table 3.11 – Purification of thyroid peroxidase (TPO) from hog thyroid glands.

1. Homogenize 3 kg of porcine thyroid glands in 1.15% KCl and 0.1 mM KI.
2. Centrifuge and save the pellet.
3. Wash the pellet with 1.15% KCl and 0.1 mM KI Recentrifuge.
4. Suspend the pellet in 0.25 M sucrose and 0.1 mM KI.
5. Homogenize and filter through cheesecloth. Save the filtrate.
6. Centrifuge the filtrate and save the pellet.
7. Repeat steps 4–6 two more times.
8. Homogenize the pellet in 1 M NaCl, 10 mM Tris buffer (pH 7.4) and 0.1 mM KI.
9. Centrifuge and save pellet.
10. Wash the pellet twice with 0.1 mM KI and centrifuge. Save the pellet.
11. Suspend the pellet in 20 mM Tris buffer (pH 8.0) containing 5 mM $CaCl_2$, 0.1 mM KI and trypsin.
12. Terminate the enzymatic reaction with soybean trypsin inhibitor.
13. Centrifuge and save supernatant.
14. Load the supernatant on a DEAE-cellulose column.
15. Wash the column with 15 mM Tris buffer (pH 7.4) containing 25 mM KCl and 0.1 mM KI.
16. Elute the enzyme with a linear KCl gradient from 25 to 225 mM KCl containing 0.1 mM KI and 15 mM Tris buffer (pH 7.4).
17. Concentrate the pooled fractions by ultrafiltration.
18. Load the concentrated fraction on a Sephacryl S-200 column.
19. Elute the enzyme with 50 mM potassium phosphate buffer (pH 7.2).

(A yield of about 35 mg of TPO was obtained; approximately 30% overall recovery.)

Reference: Neary *et al.* (1984).

Even though the strategies used involve several steps, some impressive enzyme recovery yields and purities have been achieved. The recent introduction of monoclonal antibody-assisted immunoaffinity chromatography, such as has been applied to thyroid peroxidase (Nakagawa *et al.* 1985) should lead to larger quantities of very pure enzyme preparations. As with any protein product, care in sample handling must be applied. For example, although horseradish peroxidase was effectively purified by reverse-phase high performance liquid chromatography (HPLC), the activity of the enzyme was lost during the process, indicating protein denaturization by the organic solvents used (Luiken *et al.* 1984).

Fortunately, some of the haloperoxidases are commercially available, in a variety of forms, as noted in Table 3.12.

Table 3.12 – Commercially available haloperoxidases.

Chloroperoxidase (from Caldariomyces fumago)
Sigma Chemical Co. (partially purified lyophilized powder and purified suspension)
Horseradish Peroxidase (from horseradish root)
Sigma Chemical Co. (lyophilized powder and suspension; lyophilized powders also available in type VII, VIII, and IX isoenzymes)
Calbiochem-Behring (lyophilized solid)
Lactoperoxidase (from bovine milk)
Sigma Chemical Co. (lyophilized powder and suspension)
Calbiochem-Behring (lyophilized solid)
Myeloperoxidase (from human white blood cells)
Calbiochem-Behring (pure)

Calbiochem-Behring, San Diego, CA 92112.
Sigma Chemical Co., St. Louis, MO 63172.

COMPENDIUM OF *IN VITRO* HALOPEROXIDASE REACTIONS

Enzymatic reactions often have three general characteristics: 1) reversibility, 2) high product selectivity, and 3) narrow substrate range. Haloperoxidases possess none of these. In Chapter 4 we will explain why these enzymatic reactions are not reversible, and we will present evidence to explain low product selectivity. However, it is because haloperoxidases have such a broad substrate range that the wide variety of *in vitro* reactions presented here is possible.

1. Reaction with alkenes

Over the years haloperoxidases have been shown to catalyze reactions with a wide range of alkene substrates to form alpha,beta-halohydrins (Fig. 3.1). In a systematic study of gaseous alkenes, it was demonstrated that isolated (e.g., propylene), conjugated (e.g., butadiene), and cumulative (e.g., allene) carbon–carbon double bonds were reactive (Geigert *et al.* 1983b). The molecular size of the alkene substrate seems of little importance since steroids and bicyclic compounds are also reactive.

Functional groups present on the alkene can lead to products other than halohydrins. Haloperoxidases form haloacetones from unsaturated carboxylic acids owing to the presence of an adjacent carboxyl anion (Fig. 3.2). Depending upon the alkene substrate, the major product can be either halo-δ-lactone or halo-γ-lactone. The haloperoxidase-catalyzed chlorination of allyl bromide presents yet another form of neighboring group participation (Fig. 3.3). In this example, the allylic bromine atom migrated to the central carbon during the enzymatic reaction. Carbon loss can also occur when certain functional groups are present. For example, when oxygen is attached to the carbon–carbon double bond, the enzymatic halogenation reaction can lead to a loss of carbon atoms in the form of formaldehyde or acetate (Fig. 3.4).

If halide anion effectively competes with hydroxyl anion for the presumed halonium ion intermediate, a dihalide product is formed instead of the expected halohydrin. The product distribution in such a reaction has been found to be dependent upon the halide ion(s) concentration in the reaction mixture (Fig. 3.5). Dihalide product formation is favored over halohydrin formation when increasing levels of halide ion (mM $\rightarrow$ M levels) are used in the reaction mixture. Mixtures of halide ions present in the reaction can result in heterogeneous dihalide product formation. A major impact of these observations toward the possible origin of certain marine metabolites is discussed in Chapter 5.

A sample protocol for the haloperoxidase-catalyzed halogenation of alkenes is presented in Table 3.13. The potential commercial importance of converting alkenes to halohydrins and then subsequently to epoxides will be discussed in Chapter 7.

Ref.

$$\underset{\text{ethylene}}{\mathrm{CH_2{=}CH_2}} \xrightarrow[\mathrm{Cl^-,\,H_2O_2}]{\text{chloroperoxidase}} \mathrm{\overset{OH}{\overset{|}{C}}H_2{-}\overset{Cl}{\overset{|}{C}}H_2}$$ (1)

$$\underset{\text{propylene}}{\mathrm{CH_3{-}CH{=}CH_2}} \xrightarrow[\mathrm{I^-,\,H_2O_2}]{\text{chloroperoxidase}} \mathrm{CH_3{-}\overset{OH}{\overset{|}{C}}H{-}\overset{I}{\overset{|}{C}}H_2 + CH_3{-}\overset{I}{\overset{|}{C}}H{-}\overset{OH}{\overset{|}{C}}H_2}\ (90{:}10)$$ (1)

$$\underset{\text{1,3-butadiene}}{\mathrm{CH_2{=}CH{-}CH{=}CH_2}} \xrightarrow[\mathrm{Br^-,\,H_2O_2}]{\text{lactoperoxidase}} \mathrm{CH_2{=}CH{-}\overset{OH}{\overset{|}{C}}H{-}\overset{Br}{\overset{|}{C}}H_2 + CH_2{=}CH{-}\overset{Br}{\overset{|}{C}}H{-}\overset{OH}{\overset{|}{C}}H_2 + \overset{Br}{\overset{|}{C}}H_2{-}\overset{OH}{\overset{|}{C}}H{-}\overset{OH}{\overset{|}{C}}H{-}\overset{Br}{\overset{|}{C}}H_2}\ (90{:}10{:}\text{trace})$$ (1)

$$\underset{\text{allene}}{\mathrm{CH_2{=}C{=}CH_2}} \xrightarrow[\mathrm{Br^-,\,H_2O_2}]{\text{chloroperoxidase}} \mathrm{CH_2{=}\overset{Br}{\overset{|}{C}}{-}\overset{OH}{\overset{|}{C}}H_2}$$ (1)

$$\underset{\text{butene-2}}{\mathrm{CH_3CH{=}CHCH_3}} \xrightarrow[\mathrm{Br^-,\,H_2O_2}]{\text{chloroperoxidase}} \mathrm{CH_3\overset{OH}{\overset{|}{C}}H{-}\overset{Br}{\overset{|}{C}}HCH_3}$$ (1)

$$\underset{\text{propenylphosphonic acid}}{\mathrm{CH_3CH{=}CHPO_3H_2}} \xrightarrow[\mathrm{Cl^-,\,H_2O_2}]{\text{chloroperoxidase}} \mathrm{CH_3\overset{OH}{\overset{|}{C}}H{-}\overset{Cl}{\overset{|}{C}}HPO_3H_2}$$ (2)

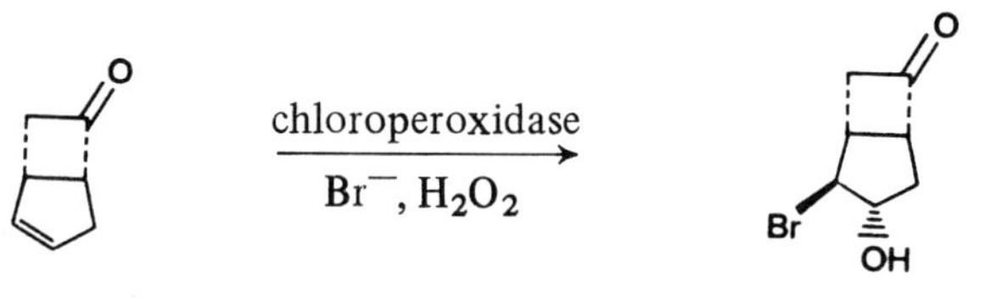

(3)

bicyclo(3.2.0)hept-2-en-6-one

Fig. 3.1 – Reaction of haloperoxidases with alkenes to form alpha, beta-halohydrins (*continued next page*).

Ref.

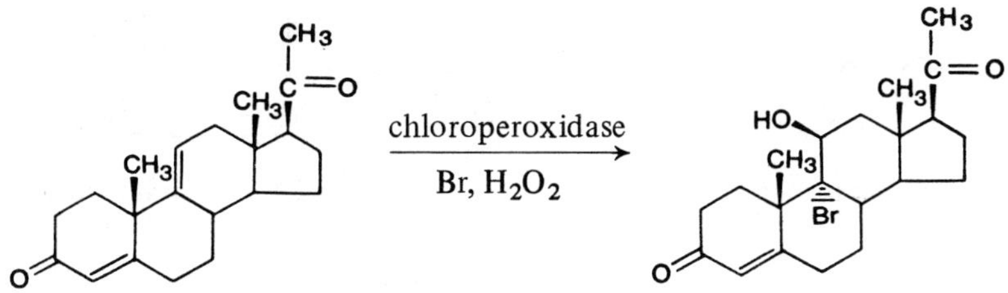

(4)

9(11)-dehydroprogesterone

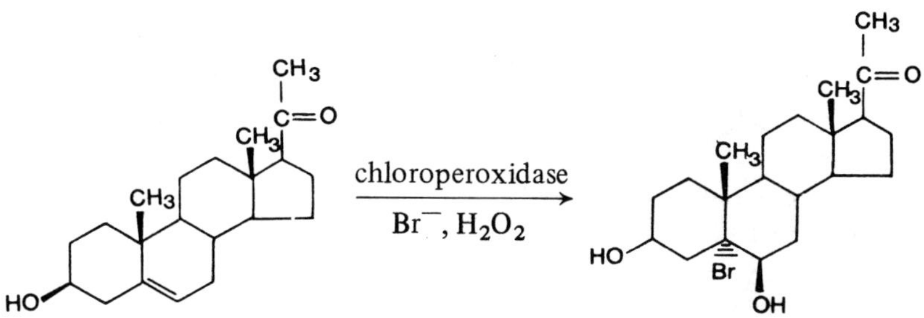

(4)

pregnenolone

References:

1. Geigert *et al.* (1984b).
2. Kollonitsch *et al.* (1970).
3. Ramakrishnan *et al.* (1983).
4. Neidleman & Levine (1968).

Fig. 3.1 – Reaction of haloperoxidases with alkenes to form alpha, beta-halohydrins.

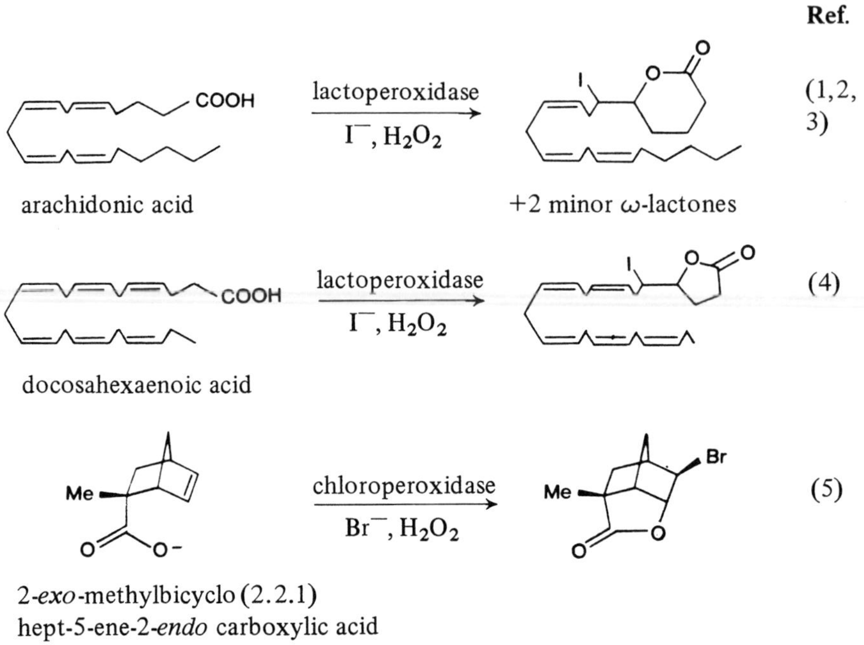

References:

1. Boeynaems & Hubbard (1980).
2. Boeynaems *et al.* (1981a).
3. Turk *et al.* (1983).
4. Boeynaems *et al.* (1981b)
5. Ramakrishnan *et al.* (1983).

Fig. 3.2 – Reaction of haloperoxidases with alkenes to form halolactones.

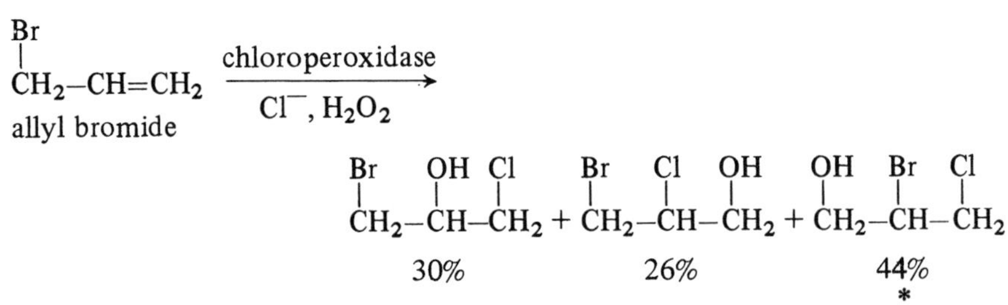

* Product formed owing to neighboring group participation.
Reference: Lee *et al.* (1983a).

Fig. 3.3 – Neighboring group participation in the chloroperoxidase-catalyzed chlorination of allyl bromide.

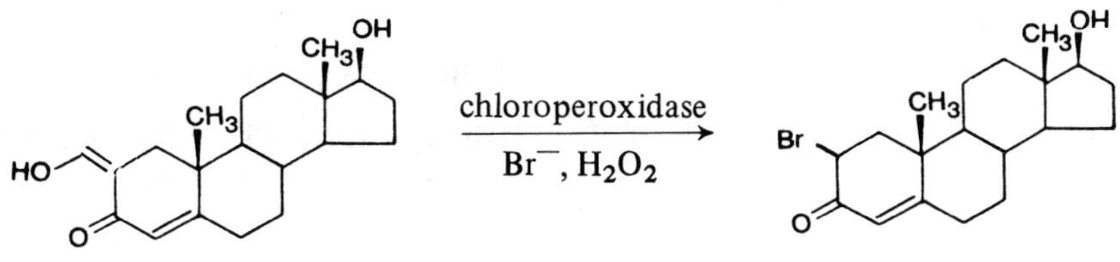

2-hydroxymethylene
testosterone

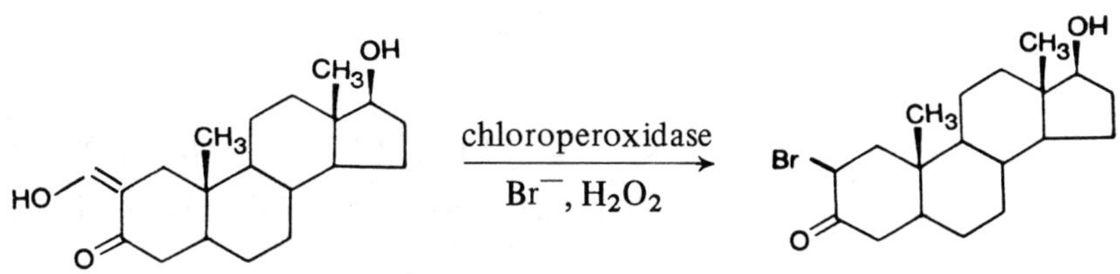

2-hydroxymethylene-17β-
hydroxy-androstan-3-one

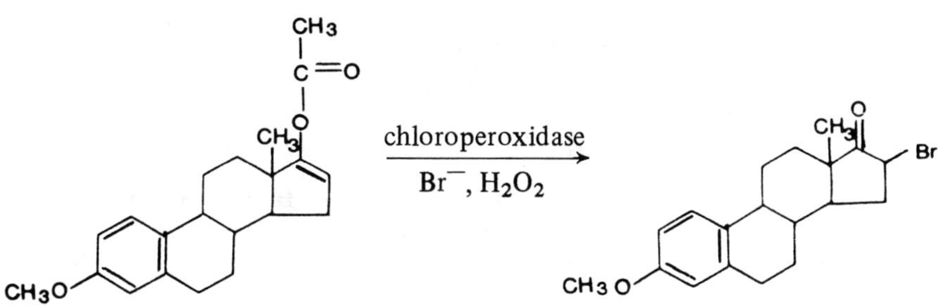

16-dehydroestradiol,
3-methylether 17-acetate

References: Neidleman & Oberc (1968), Levine *et al.* (1968).

Fig. 3.4 – Reaction of haloperoxidases with alkenes containing oxygen attached to the double bond.

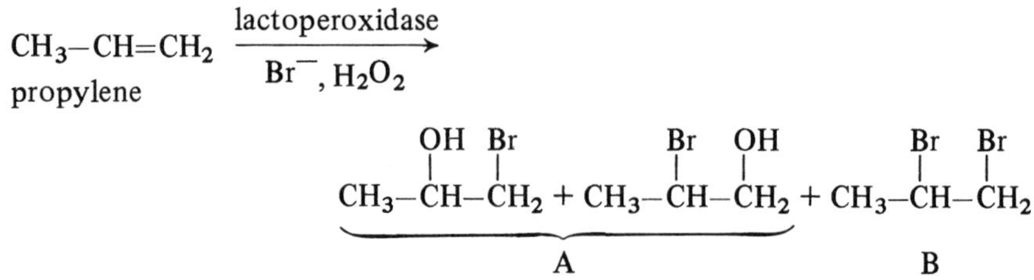

$[Br^-]$	Ratio B : A
20 mM	<0.01
2000 mM	2

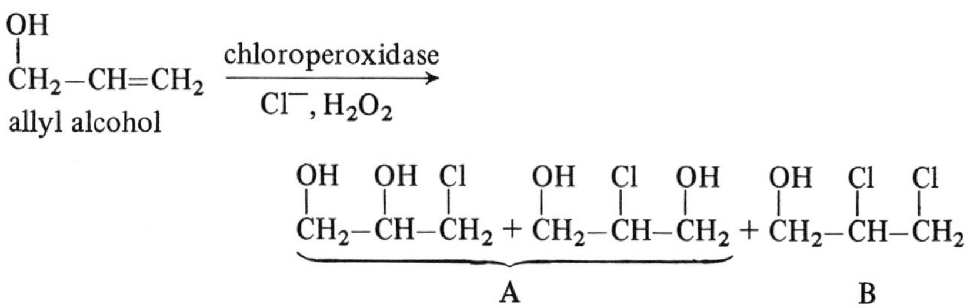

$[Cl^-]$	Ratio B : A
20 mM	0.1
2000 mM	8

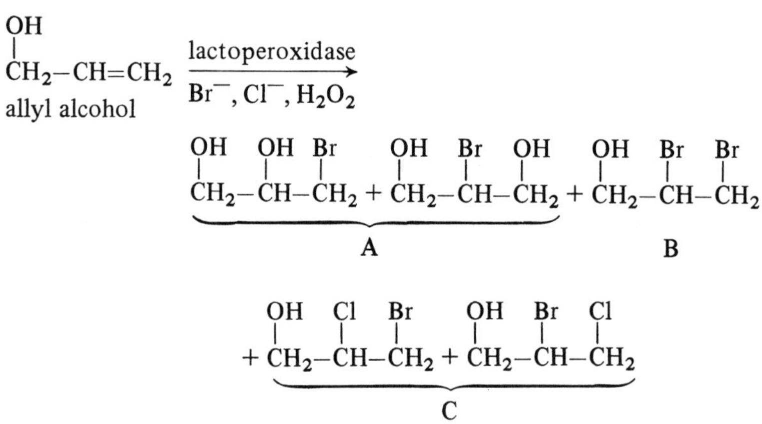

$[Br^-]$	$[Cl^-]$	Ratio C : B : A
20 mM	2000 mM	10 : 0.9 : 0.7

References: Geigert *et al.* (1983c), Neidleman & Geigert (1983a).

Fig. 3.5 – Reaction of haloperoxidases with alkenes to form alpha, beta-dihalides.

Table 3.13 – Reaction protocol for the chloroperoxidase-catalyzed bromination of allyl chloride.

Reaction:

$$\underset{\text{allyl chloride}}{Cl{-}CH_2{-}CH{=}CH_2} + Br^- + H_2O_2 \xrightarrow{\text{chloroperoxidase}}$$

$$\underset{\text{bromohydrins of allyl chloride}}{Cl{-}CH_2{-}\overset{OH}{\overset{|}{C}}H{-}\overset{Br}{\overset{|}{C}}H_2 + Cl{-}CH_2{-}\overset{Br}{\overset{|}{C}}H{-}\overset{OH}{\overset{|}{C}}H_2}$$

Materials:

Reaction vessel:	10 ml glass vial, septum sealed; magnetic stir bar and stirrer
Reaction mixture:	(4.5 ml) 0.3 M potassium phosphate buffer pH 3.0
	(0.4 ml) 0.42 M potassium bromide
	(15 μl) allyl chloride
	(0.1 ml) 3% hydrogen peroxide
Initiation Reagent:	(0.1 ml) chloroperoxidase (5×10^4 units*)

Procedure:

1. Ready the reaction mixture in the reaction vessel.
2. Initiate the bromination reaction by the addition of enzyme.
3. Assay mixture after 5 min.

Assay:

Gas chromatography; 6 foot glass column packed with Tenex-GC (80–100 mesh); column temperature maintained at 190°C isothermal; column flow set at 40 ml/min He; detection by flame ionization (FID). Direct aqueous injection (5 μl) of reaction solution.

Discussion:

The reagent levels in the reaction mixture are optimized for bromohydrin formation: a ratio of 1:2:3; H_2O_2:Br^-: allyl chloride, on a molar basis. Approximately 20% of the added bromide is converted to bromohydrins (~15 μmoles) under these conditions. Increase the Br^- level and dibromination instead of bromohydrin formation will occur. Increase the H_2O_2 level and the enzyme will be inactivated.

Additional assays that have been used to follow the reaction include HPLC (Geigert *et al.* 1981) and fused silica capillary GC/MS (Hirano *et al.* 1983).

---

*One unit of activity causes a decrease in A_{277nm} of 0.001 per minute at pH 2.8 at 25°C, using monochlorodimedone as the halogen acceptor.

2. Reaction with alkynes

With alkyne substrates, haloperoxidase-catalyzed reactions yield alpha-haloketones (Fig. 3.6). These enzymes are thus distinguished by being able to react with both carbon–carbon double bonds and triple bonds.

The product distribution in the reaction with alkynes is dependent upon the halide ion(s) concentration in the reaction mixture (Fig. 3.7). Increasing the halide ion concentration (mM→M levels) results in the formation of alpha, beta-dihalides. As with alkenes, both homogeneous dihalides and heterogeneous dihalides can be formed from alkynes, dependent upon whether a single halide ion or a mixture of halide ions is present.

The protocol for haloperoxidase-catalyzed halogenation of alkynes is identical to that used for alkenes.

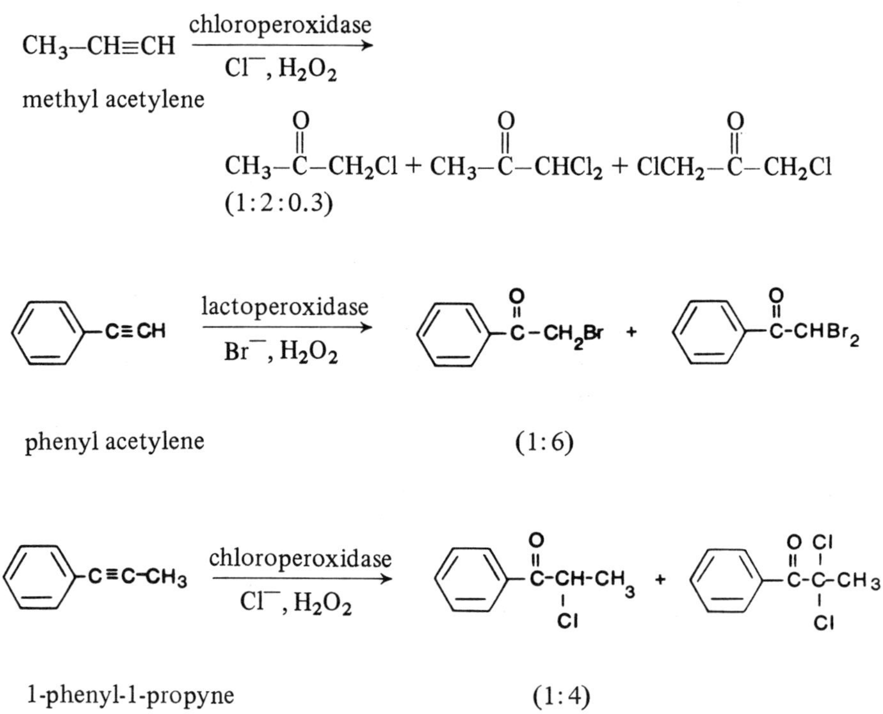

Reference: Geigert *et al.* (1983d).

Fig. 3.6 – Reaction of haloperoxidases with alkynes to form alpha-haloketones.

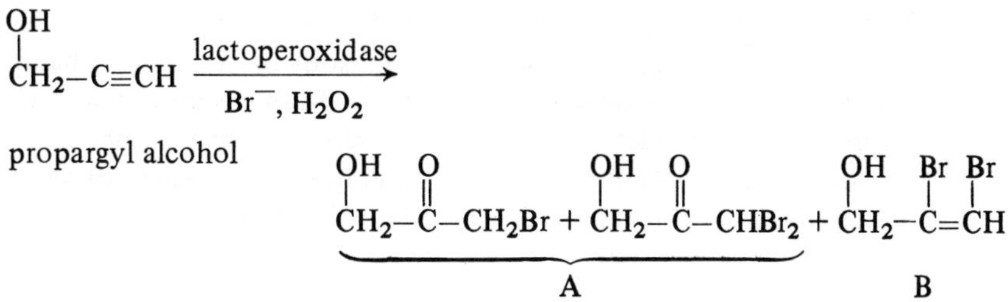

$[Br^-]$	Ratio B : A
20 mM	0.05
2000 mM	9

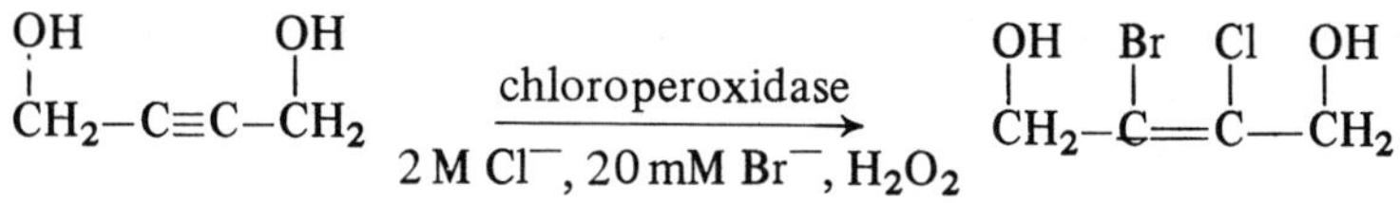

1,4-butynediol

Reference: Geigert *et al.* (1983c).

Fig. 3.7 – Reaction of haloperoxidases with alkynes to form alpha, beta-dihalides.

3. Reaction with cyclopropanes

The reaction between cyclopropanes and haloperoxidases yields alpha, gamma-halohydrins (Fig. 3.8). Relatively few cyclopropanes have been studied to date.

The protocol for haloperoxidase-catalyzed halogenation of cyclopropanes is identical to that used for alkenes.

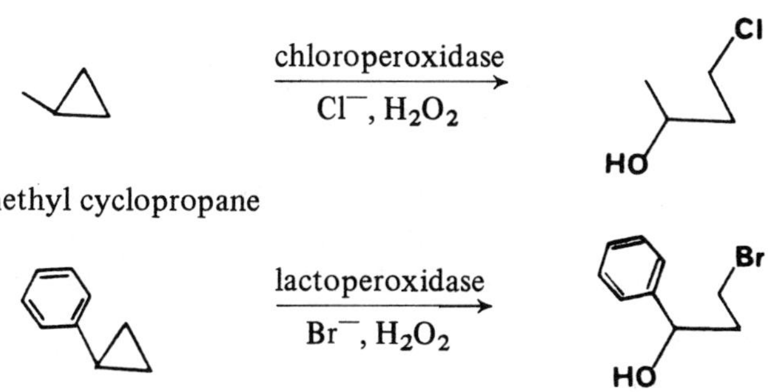

methyl cyclopropane

phenyl cyclopropane

Reference: Geigert *et al.* (1983d).

Fig. 3.8 – Reaction of haloperoxidases with cyclopropanes to form alpha, gamma-halohydrins.

4. Reaction with phenols, anilines, and other aromatics

A wide range of aromatic compounds are readily halogenated by haloperoxidases: phenols and anilines (and their respective O- and N-alkylated derivatives) and heterocyclic aromatics (Fig. 3.9). The color change in phenolic dyes upon enzymatic halogenation is a useful and simple assay system for haloperoxidases. Because the reaction can be observed visually, it has been used in screening for new enzymes (Fig. 3.10).

Since haloperoxidases are also peroxidases, they can carry out the full range of halide-independent peroxidations on aromatic compounds: dimerization, polymerization, oxygen insertion, dealkylations, etc. (Fig. 3.11). Peroxidation predominates over halogenation whenever halide ion is omitted from the reaction or whenever the halide ions become depleted during the course of the enzymatic reaction. Thus, peroxidation may be an undesirable competitive reaction.

The protocol for haloperoxidase-catalyzed halogenation of phenolic dyes is presented in Table 3.14.

Table 3.14 – Reaction protocol for the lactoperoxidase-catalyzed bromination of phenol red.

Reaction:

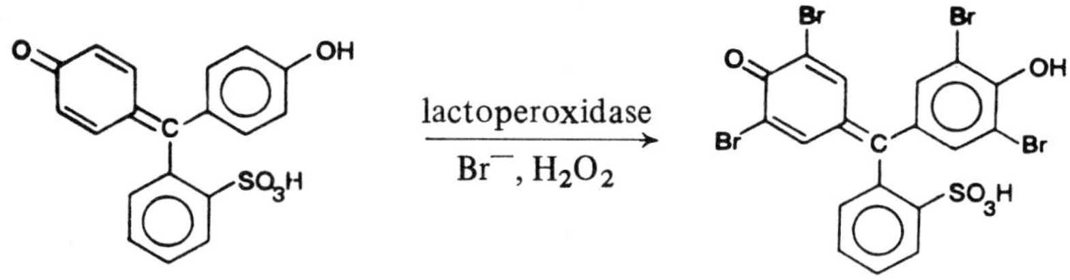

phenol red — bromophenol blue

Materials:

Reaction vessel:	10 ml glass test tube
Reaction mixture:	(5 ml) 0.1 M potassium phosphate buffer pH 6.5
	(0.05 ml) 0.42 M potassium bromide
	(0.05 ml) 0.04% phenol red
	(0.1 ml) 3% hydrogen peroxide
Initiation Reagents:	(0.1 ml) lactoperoxidase (5 units*)

Procedure:

1. Ready the reaction mixture in the reaction vessel.
2. Initiate the bromination reaction by the addition of enzyme.
3. Assay mixture after 5 min.

Assay:

Examine the color in the test tube visually (phenol red is yellow, bromophenol blue is blue at pH 6.5), or measure the increase in absorbance at 595 nm.

Discussion:

The product is actually a mixture containing some mono-, di-, and tri-brominated phenol red derivatives.

Other dyes (e.g., fluorescin) can be used in place of phenol red to provide an alternative pH for the color discrimination between substrate and product.

---

* One unit of activity forms 1 mg of purpurogallin from pyrogallol in 20 seconds at pH 6.0 at 20°C.

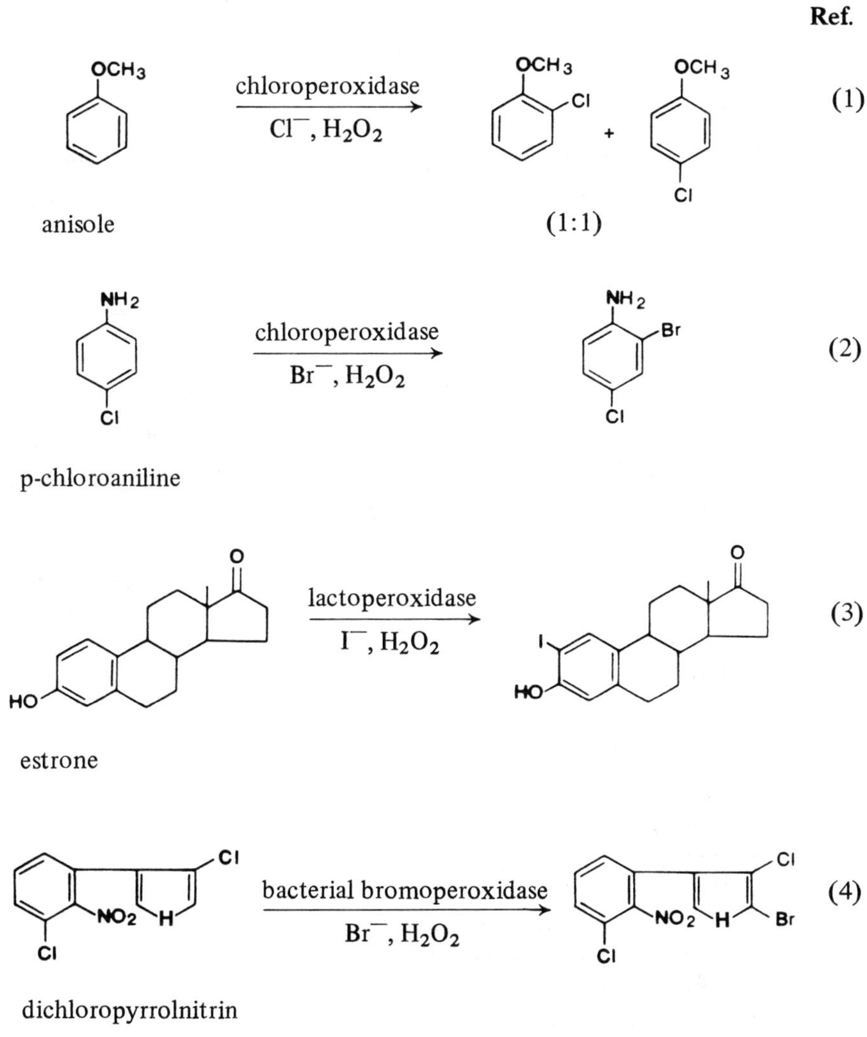

Fig. 3.9 – Reaction of Haloperoxidases with Phenols, Anilines, and Other Aromatics

Ref.

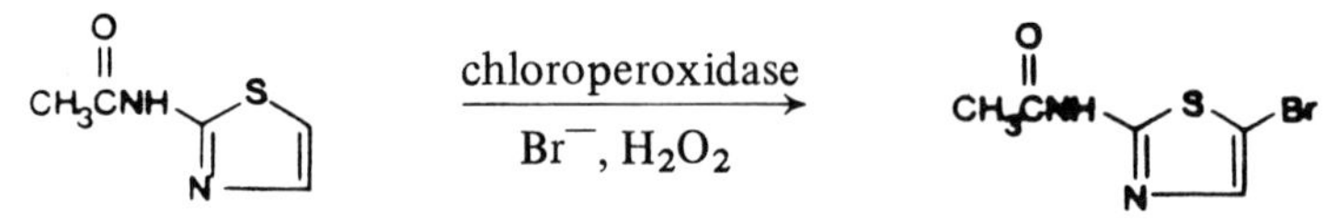

(5)

2-acetamidothiazole

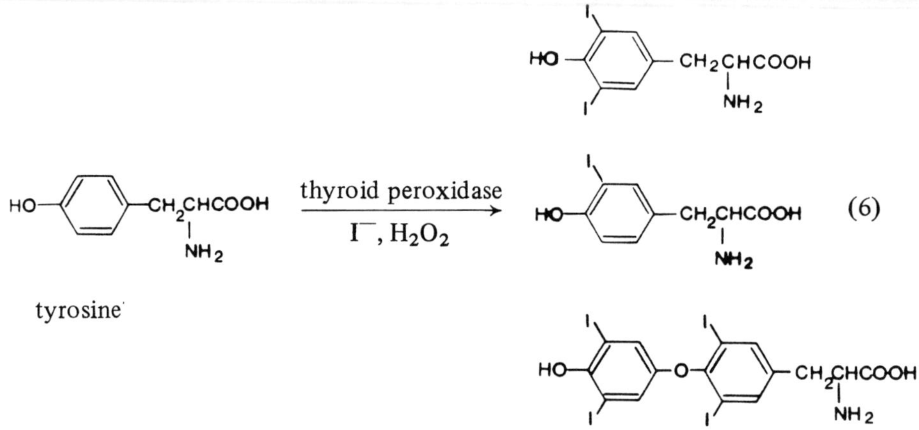

(6)

tyrosine

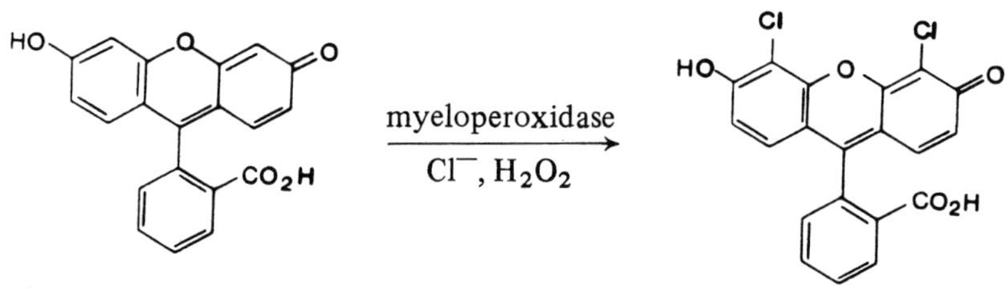

(7)

fluorescein

References:

1. Jerina *et al.* (1968).
2. Corbett *et al.* (1980).
3. Matkovics *et al.* (1971).
4. van Pee & Lingens (1984).
5. Neidleman *et al.* (1969).
6. Nunez (1984).
7. Hurst *et al.* (1984).

Fig. 3.9 – Reaction of Haloperoxidases with Phenols, Anilines, and Other Aromatics

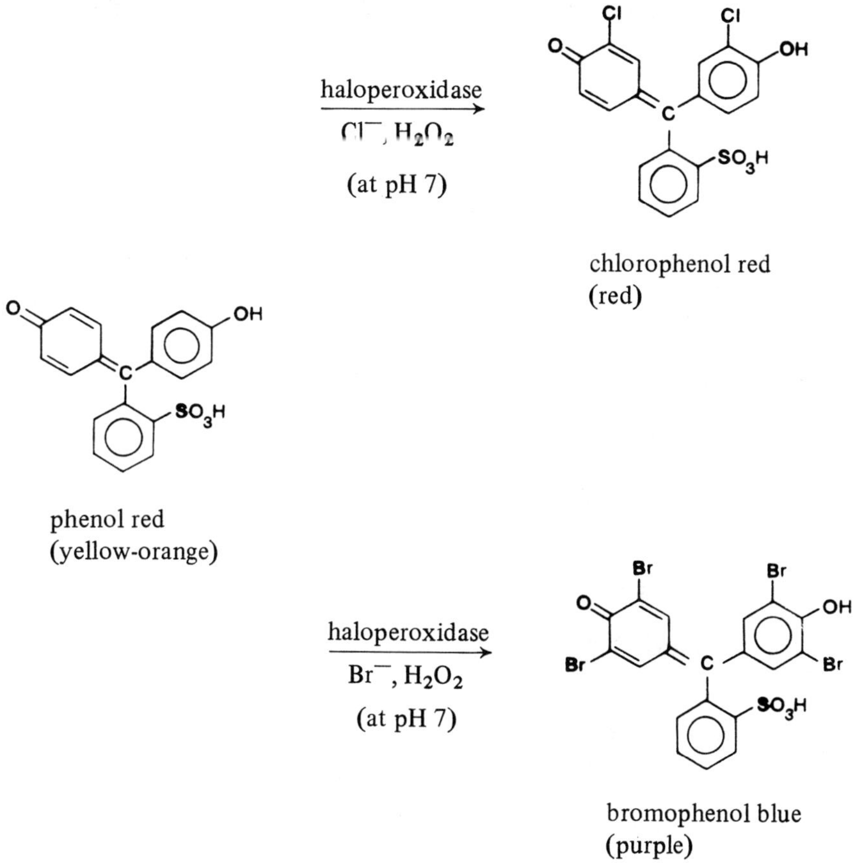

References: Hunter *et al.* (1985b), Loo *et al.* (1964).

Fig. 3.10 – Color change in phenolic dyes upon enzymatic halogenation.

Ref.

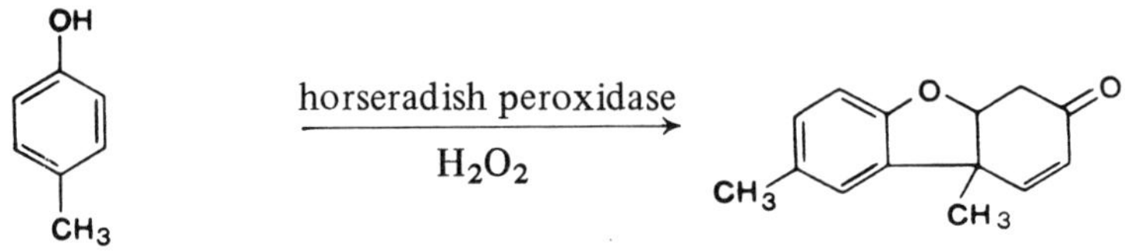

(1)

p-cresol

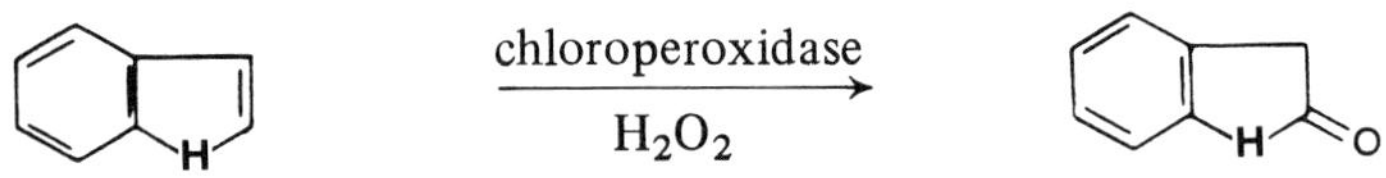

(2)

indole

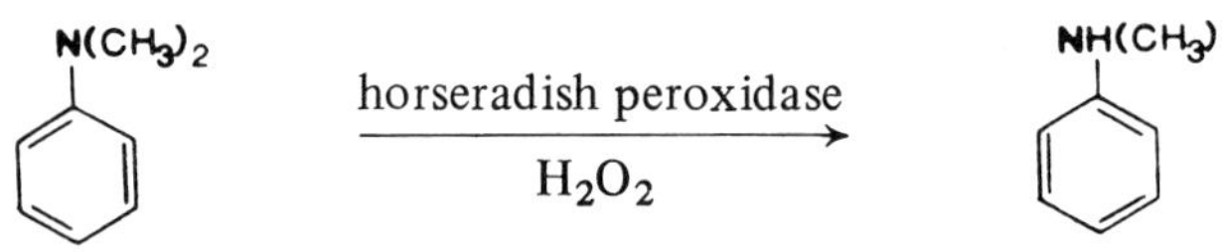

(3)

dimethylaniline

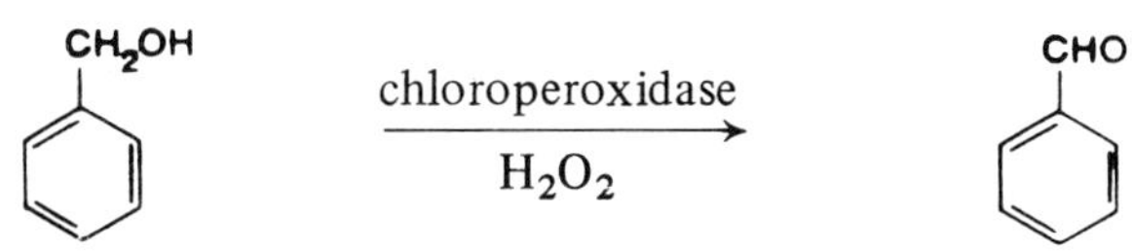

(4)

benzyl alcohol

References:

1. Hewson & Dunfold (1976).
2. Corbett & Chipko (1979).
3. Kedderis & Hollenberg (1984).
4. Geigert *et al.* (1983e).

Fig. 3.11 – Halide-independent peroxidation of aromatics by haloperoxidases.

5. Reaction with beta-diketones and beta-ketoacids

A wide range of beta-diketones have been halogenated by haloperoxidases (Fig. 3.12). Reactivity appears to be a function of percent enol content of the substrate in the reaction mixture. A simple ketone like 2-heptanone, which has little enol content, in unreactive; while a beta-diketone like monochlorodimedone (MCD), which is predominantly in the enol form in solution, is a very reactive substrate. In fact, because MCD has this high reactivity and because halogenation dramatically decreases its UV absorbance, MCD is the substrate of choice for quantitating haloperoxidase activity. The beta-diketone substrates can range from monocyclic compounds to polycyclic steroids. Beta-ketoacids readily decarboxylate upon halogenation with haloperoxidases (Fig. 3.13).

The protocol for haloperoxidase-catalyzed halogenation of beta-diketones is presented in Table 3.15.

Table 3.15 – Reaction protocol for the chloroperoxidase-catalyzed chlorination of monochlorodimedone (MCD).

Reaction:

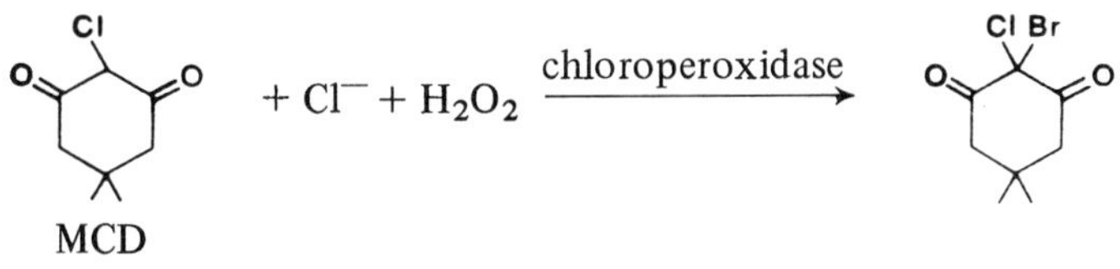

Materials:

Reaction vessel:	5 ml glass tube
Reaction mixture:	(2 ml) 0.2 M potassium phosphate buffer pH 3.0
	(0.1 ml) 50 mM potassium chloride
	(0.1 ml) 2 mM MCD
	(0.1 ml) 10 mM hydrogen peroxide
Initiation reagent:	(10 μl) chloroperoxidase

Procedure:

1. Ready the reaction mixture in the reaction vessel.
2. Initiate the chlorination reaction by the addition of enzyme.
3. Assay mixture at a fixed time point or continuously (i.e., run the reaction in a UV quartz cuvette).

Assay:

Monitor the decrease in absorbance at 292 nm, corresponding to the consumption of MCD (MCD has an ϵ value of 20 000 at this wavelength, while the chlorinated product has an ϵ value of <100).

Discussion:

The MCD reaction is the standard assay used to assign the activity of a haloperoxidase enzyme. For bromoperoxidases, bromide ion is used in place of chloride ion.

Ref.

$$CH_3(CH_2)_4\overset{\overset{O}{\|}}{C}CH_3 \xrightarrow[Br^-,\ H_2O_2]{\text{bromoperoxidase}} \text{no reaction}$$ (1)

2-heptanone
(not a β-diketone)

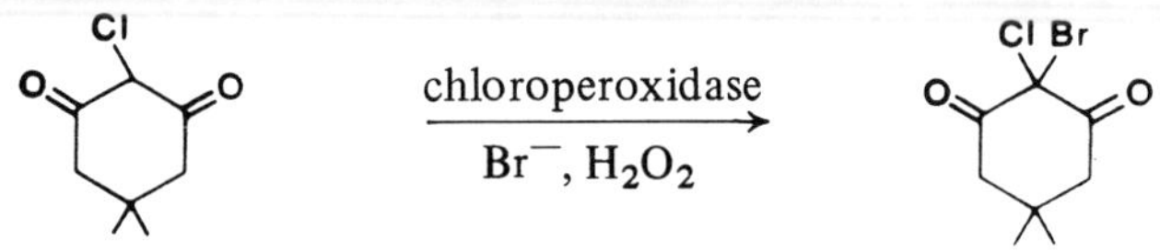

(2, 3)

monochlorodimedone
(MCD)

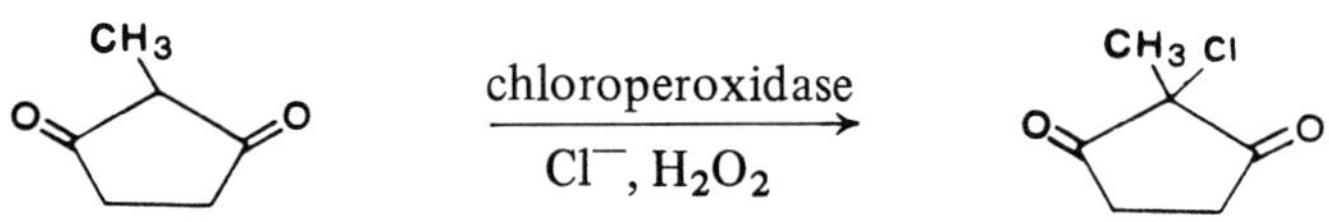

(4)

2-methylcyclopentane-
1,3-dione

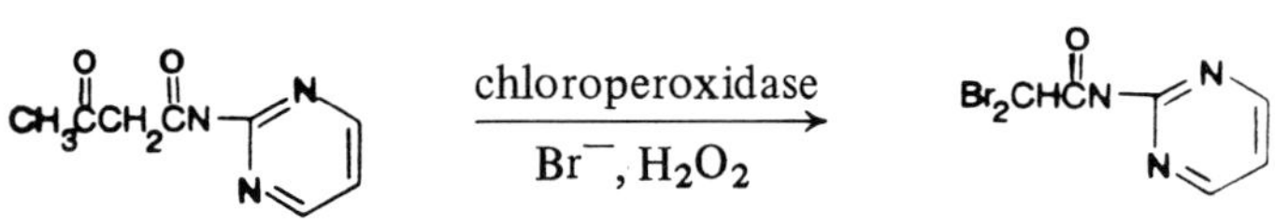

(5)

2-acetoacetamidopyrimidine

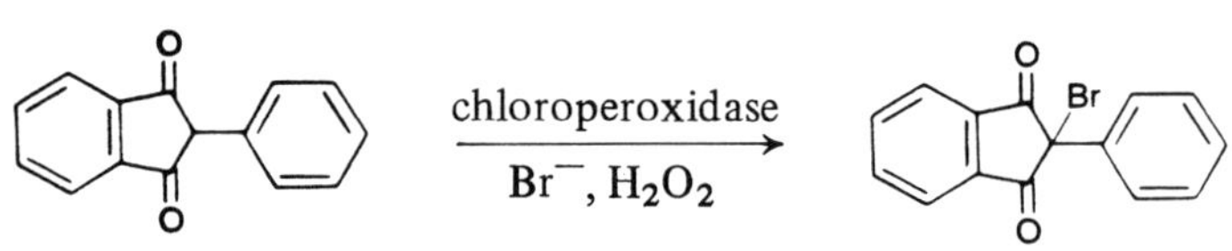

(5)

2-phenyl-1,3-indandione

Fig. 3.12 – Reaction of haloperoxidases with β-diketones
(*continued next page*).

Ref.

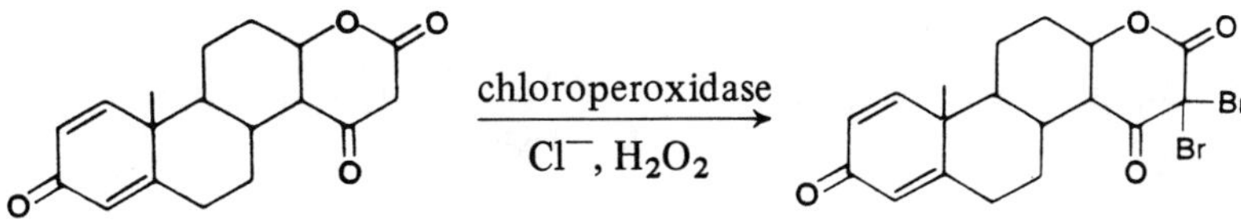

(6)

15-keto-1-
dehydrotestololactone

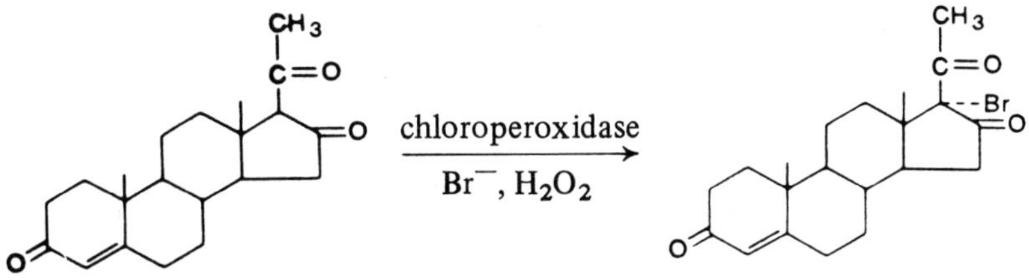

(6)

16-ketoprogesterone

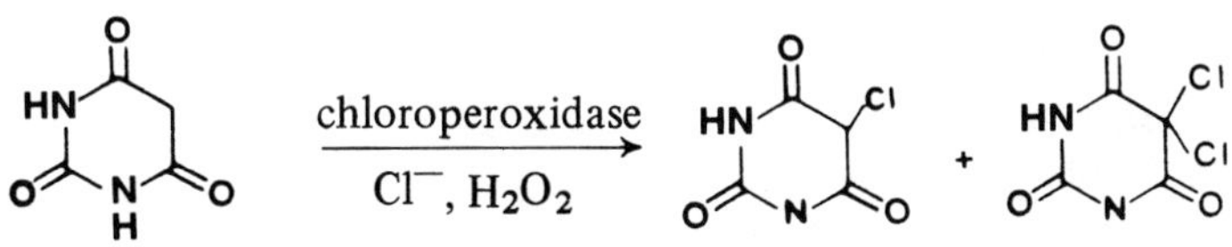

(7)

barbituric acid

References:

1. Beissner *et al.* (1981).
2. Martyn *et al.* (1981).
3. Libby *et al.* (1982).
4. Ramakrishnan *et al.* (1983).
5. Neidleman (1975).
6. Neidleman *et al.* (1966).
7. Franssen & van der Plas (1984).

Fig. 3.12 – Reaction of haloperoxidases with β-diketones.

Ref.

$$HO_2CCH_2CH_2\overset{\overset{\displaystyle O}{\|}}{C}CH_2CO_2H \xrightarrow[\text{Cl, } H_2O_2]{\text{chloroperoxidase}} HO_2CCH_2CH_2\overset{\overset{\displaystyle O}{\|}}{C}CH_2Cl + CO_2 \qquad (1)$$

3-oxoadipic acid

$$CH_3\overset{\overset{\displaystyle O}{\|}}{C}CH_2CO_2H \xrightarrow[Br^-,\, H_2O_2]{\text{bromoperoxidase}} CH_3\overset{\overset{\displaystyle O}{\|}}{C}CH_2Br + CO_2 \qquad (2)$$

acetoacetic acid

$$CH_3(CH_2)_4\overset{\overset{\displaystyle O}{\|}}{C}CH_2CO_2H \xrightarrow[Br^-,\, H_2O_2]{\text{bromoperoxidase}} CH_3(CH_2)_4\overset{\overset{\displaystyle O}{\|}}{C}CHBr_2 + CO_2 \qquad (3)$$

3-oxooctanic acid

References:

1. Shaw & Hager (1959).
2. Theiler *et al.* (1978).
3. Beissner *et al.* (1981).

Fig. 3.13 – Reaction of haloperoxidases with β-ketoacids.

6. Reaction with nitrogen atoms

Haloamines are the primary products formed when haloperoxidases react with amine compounds, but because many of these haloamines are unstable, they rapidly deaminate/decarboxylate, freeing the halogen (Fig. 3.14). This reactivity with amines serves as part of the natural defense system in mammals, as will be described in Chapter 6.

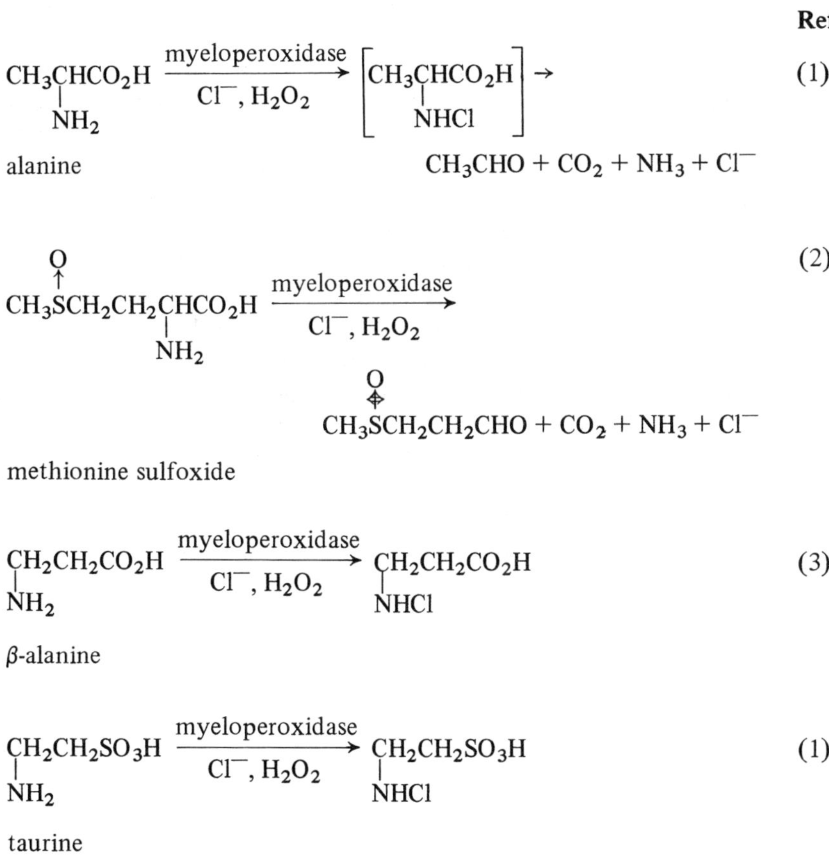

References:

1. Grisham *et al.* (1984).
2. Tsan (1982a).
3. Selvaraj *et al.* (1978).

Fig. 3.14 – Reaction of haloperoxidases with nitrogen atoms.

7. Reaction with sulfur atoms

Thiols (–SH) are oxidized by haloperoxidases to yield the sulfenyl halide (–SX). These in turn can react rapidly with excess thiol compounds to yield a disulfide (–S–S–) or with a nucleophile, such as hydroxyl anion, to yield a sulfonic acid ($-SO_3H$) derivative (Fig. 3.15). Disulfides, like cystine, and dialkyl sulfides, like methionine, are also oxidized by haloperoxidases in the presence of halide ions yielding sulfoxides. Acting as a peroxidase in the absence of halide ion, the chloroperoxidase from *Caldariomyces fumago* oxidizes dimethyl sulfoxide (DMSO) to its sulfone (Fig. 3.16). None of the other haloperoxidases appear capable of catalyzing this reaction. Other sulfoxides, like methionine sulfoxide, are apparently not oxidized by this chloroperoxidase, even in the presence of halide ion. Another demonstrated peroxidation reaction is the dismutation of dithionite ($S_2O_4^{2-}$) to thiosulfate ($S_2O_3^{2-}$) and hydrogensulfite (HSO_3^-), with lactoperoxidase (Ohlsson 1984). Thiocyanate (SCN^-) is a pseudo-halide and in the presence of hydrogen peroxide, most haloperoxidases will oxidize thiocyanate to hypothiocyanous acid (HOSCN).

A practical application of the halogenation of thiols can be illustrated by early literature on the effect of antithyroid compounds such as 2-mercaptobenzothiazole (which contains a –SH group) to decrease the ratio of 7-chlortetracycline (CTC): tetracycline in the fermentation process for production of these antibiotics. The effects of such antithyroid compounds on CTC production can be attributed to the fact that they are better substrates for the chlorinating enzyme involved in CTC production than the ‘natural’ substrate (Neidleman 1975).

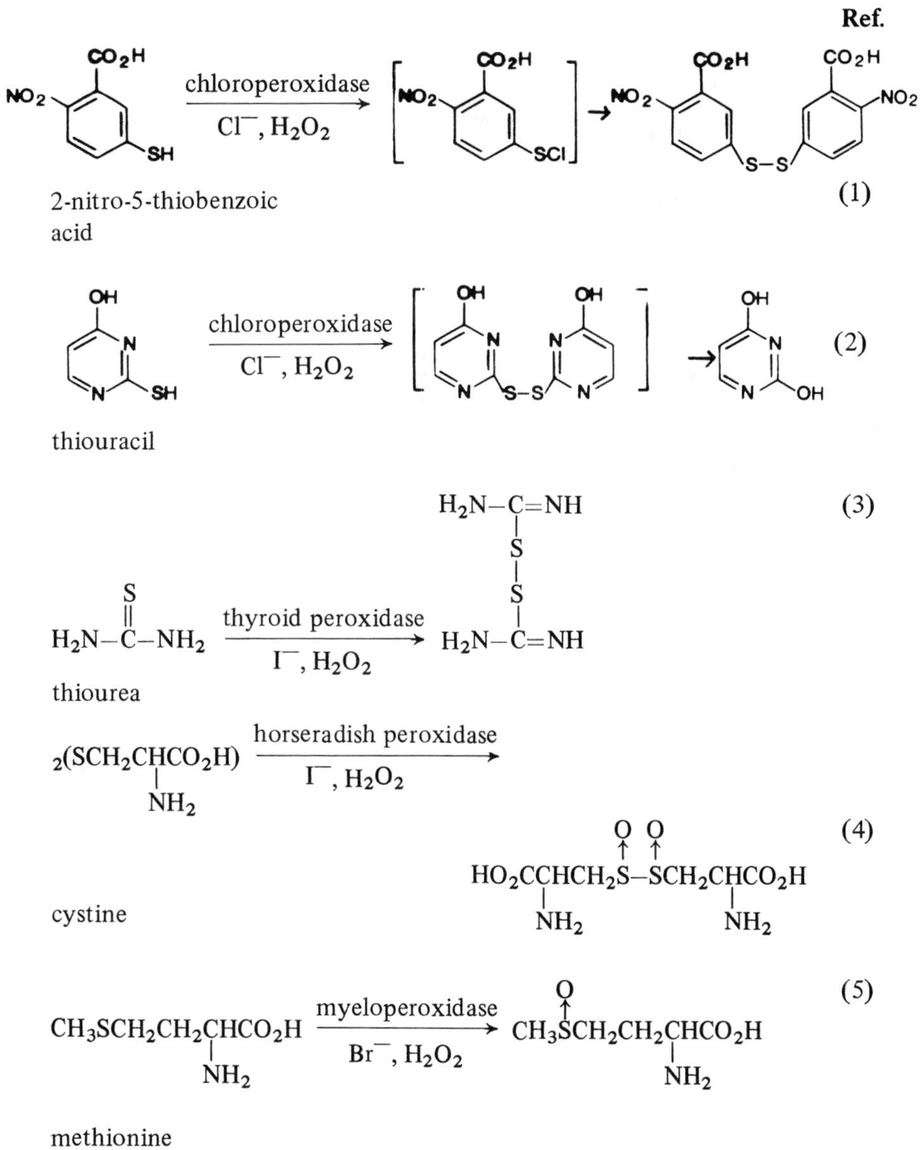

References:

1. Silverstein & Hager (1974).
2. Morris & Hager (1966a).
3. Davidson *et al.* (1979).
4. Stelmaszynska & Zgliczynski (1963).
5. Tsan (1982b).

Fig. 3.15 – Reaction of haloperoxidases with sulfur atoms.

Ref.

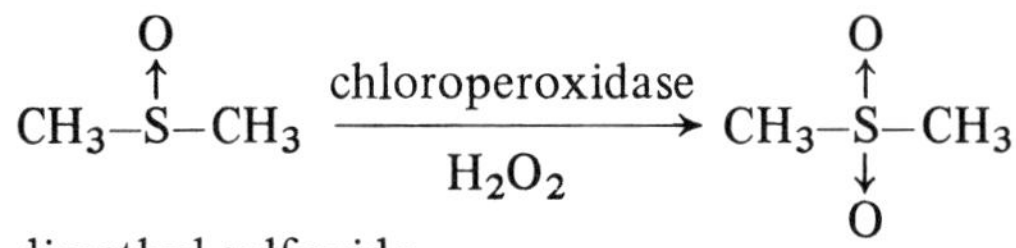

(1)

dimethyl sulfoxide
(DMSO)

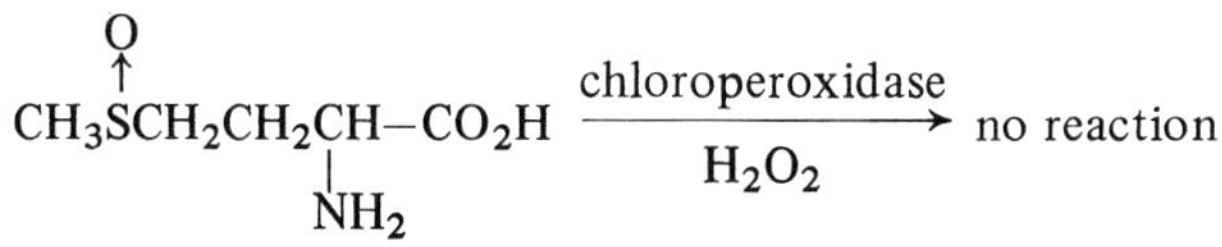

(2)

methionine sulfoxide

References:

1. Geigert *et al.* (1983f).
2. Libby *et al.* (1982).

Fig. 3.16 – Halide-independent reaction of chloroperoxidase with sulfur atoms.

8. Reaction with inorganic halogens other than halides

The chloroperoxidase from *C. fumago* has unique reactivities toward certain inorganic molecules (Fig. 3.17). Of all the haloperoxidases, it alone can oxidize iodine to iodate. Although both chloroperoxidase and horseradish peroxidase can convert chlorite to chlorine dioxide, only chloroperoxidase can dismutate chlorine dioxide to chlorate.

Ref.

$$\underset{\text{iodine}}{I_2} \xrightarrow[H_2O_2]{\text{chloroperoxidase}} \underset{\text{iodate}}{IO_3^-}$$ (1)

$$\underset{\text{chlorite}}{ClO_2^-} \xrightarrow[\text{horseradish peroxidase}]{\text{chloroperoxidase}} \underset{\text{chlorine dioxide}}{ClO_2}$$ (2, 3)

$$\underset{\text{chlorine dioxide}}{ClO_2} \xrightarrow{\text{chloroperoxidase}} \underset{\text{chlorate}}{ClO_3^-}$$

References:

1. Thomas & Hager (1968).
2. Shahangian & Hager (1981).
3. Hewson & Hager (1979).

Fig. 3.17 – Reaction of haloperoxidases with inorganic halogens other than halides.

IN VIVO HALOGENATION

In one *in vivo* study, it has been clearly demonstrated that thyroid peroxidase is responsible for the iodination and coupling of the amino acid tyrosine to yield the thyroid hormone thyroxine, structures shown in Fig. 2.7 (Nunez 1984). From such a study, as well as realizing that haloperoxidases have a wide distribution throughout nature and a broad substrate specificity, it is tempting to say that haloperoxidases are responsible for all halometabolites discovered to date, but other *in vivo* studies suggest that this is not the case.

The biosynthesis by *Ochromonas dancia* of the chlorosulfolipids listed in Fig. 2.4 has been investigated using labeled precursors. These studies have revealed that chlorination occurs after the C_{22}-carbon backbone has been completed (Elovson 1974). Chloroperoxidases would not seem to be involved since no haloperoxidase has been demonstrated to halogenate an unactivated carbon site *in vitro*.

In vivo formation of bromometabolites in place of the 'normal' chlorometabolites can be readily accomplished by merely replacing chloride ion with bromide ion in the growth medium. The list of microorganisms that have successfully yielded bromo analogs in this way is presented in Table 3.16. While this *in vivo* effect could be accomplished by haloperoxidases, an appropriate enzyme has been isolated from only two of these microorganisms, *Caldariomyces fumago* and *Pseudomonas aureofaciens.*

The biosynthesis of fluorometabolites remains a mystery since, as we will discuss in the next chapter, haloperoxidases cannot directly cause fluoride oxidation.

Needless to say, additional studies are needed to better understand how nature yields such a diversity of halometabolites. Novel halogenating enzymes must exist.

Table 3.16 – Biosynthesized bromo analogues.

Producing microorganisms	Bromo analog of	Reference
Caldariomyces fumago	caldariomycin	Patterson *et al.* (1967)
Penicillium griseofulvum	griseofulvin	Neidleman (1975)
Streptomyces venezuelae	chloramphenicol	Neidleman (1975)
Streptomyces aureofaciens	chlortetracycline	Neidleman (1975)
Pseudomonas aureofaciens	chloropyrrolnitrin	van Pee *et al.* (1983)
Actinosporangium vitaminophilum	chloropyrrolomycin	Ezaki *et al.* (1983b)
Ochromonas danica	chlorosulfolipids	Neidleman (1975)

4

The halogenating enzymes: properties

INTRODUCTION

Peroxidases are enzymes that catalyze the oxidation of a wide spectrum of substrates with hydrogen peroxide or other hydroperoxides. Haloperoxidases are the subgroup of peroxidases that can catalyze the oxidation of halide ions (I^-, Br^-, and Cl^-, but not F^-). Thus, every haloperoxidase ia a peroxidase, but the reverse is not so.

Much progress was made in the 1950s and 1960s in understanding the properties of the enzymatic intermediates and the kinetics of peroxidatic activity. This progress was due to the talented researchers involved in the investigations (for example, the pioneering work of B. Chance and P. George) and the availability of the enzymes in highly purified form. The stability of the enzyme intermediates (half-life of hours for Compound II, minutes for Compound I) and the intense, environmentally-sensitive Soret absorbance band at about 410 nm greatly facilitated such studies. However, in terms of the structure at the active site of peroxidases, little is known for certain. Despite the recent application of advance analytical techniques (for example, resonant Raman scattering, electron paramagnetic resonance, and nuclear magnetic resonance), several areas, such as the ligand in the sixth coordination position of the heme iron and the nature of the halogenating intermediate for haloperoxidases, remain topics of controversy. It has been nine years since the last exhaustive review of haloperoxidases was published (Morrison & Schonbaum 1976). It is our hope that the up-to-date information presented in this chapter will encourage additional research on these enzymes. Haloperoxidases offer fascinating problems in trying to decipher how subtle changes in physicochemical structure can lead to drastic changes in catalytic properties.

THE COMPONENTS

As will be seen in this section, it is difficult to definitively describe the basic components of this class of enzymes. In general, it can be said that haloperoxidases consist of heme and glycoprotein. The structure and intimate interaction between these two components imparts to each haloperoxidase a unique set of catalytic properties.

The heme prosthetic group at the active site of the enzyme has been studied extensively and has been shown to have a dual role in the catalytic function of haloperoxidases:

1. The metal (i.e., iron(III)) activates the heterolytic cleavage of H_2O_2 and stores one oxidizing equivalent, and
2. the porphyrin regulates the oxidation-reduction potential and stores one oxidizing equivalent.

These roles will receive further discussion in the section on reaction mechanisms.

While most haloperoxidases contain ferriprotoporphyrin IX (see Fig. 4.1) as the heme component, Table 4.1 shows that myeloperoxidase is a known exception. Myeloperoxidase has an additional two-fold distinction. First, the shift of the Soret absorbance band from 428 nm to 472 nm (a shift of 44 nm) that occurs upon reduction with dithionite is the largest shift reported for a heme-containing protein, and second, the enzyme contains two hemes per molecule. There is some question as to whether both hemes are structurally equivalent, but it has been proposed that the heme on this enzyme contains a formyl group (–CHO) replacing the methyl group ($-CH_3$) at position 8 in Fig. 4.1.

Table 4.1 – Haloperoxidase prosthetic groups.

Enzyme	Ferriporphyrin prosthetic group	Number per enzyme	Reference
myeloperoxidase	not IX	2	Schultz (1980)
eosinophil peroxidase	IX	1	Bolscher *et al.* (1984a)
chloroperoxidase (*C. fumago*)	IX	1	Morris & Hager (1966b)
lactoperoxidase	IX	1	Sievers (1979)
bromoperoxidase (*P. capitatus*)	IX	1	Manthey & Hager (1981)
bromoperoxidase (*P. aureofaciens*)	IX	1	van Pee & Lingens (1985)
ovoperoxidase	IX	1	Deits *et al.* (1984)
thyroid peroxidase	IX	1	Ohtaki *et al.* (1985)
horseradish peroxidase	IX	1	Dunford & Stillman (1976)

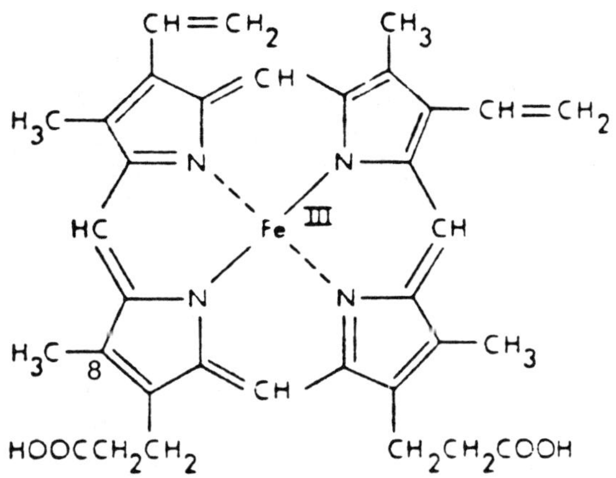

Fig. 4.1 – The heme component of many haloperoxidases – ferriprotoporphyrin IX.

To further confuse the issue, it has now been shown that heme is not even necessary in the active site. Non-heme lactoperoxidase has been discovered in bovine milk (Dumontet & Rousset 1983). Both lactoperoxidase (i.e., ferriprotoporphyrin IX-containing) and non-heme lactoperoxidase have identical apparent molecular weights, protein molar absorption coefficients at 280 nm, proteolytic peptide maps, carbohydrate content, and immunological cross-reactivity. However, lactoperoxidase possesses a Soret band at 412 nm, while non-heme lactoperoxidase was devoid of all light-absorption properties in the 300–500 nm region. Non-heme lactoperoxidase is active, but with a specific activity of less than 1% of that found for lactoperoxidase, as measured by the oxidation of I^- to I_3^-. Additional heme-minus enzymes include a non-heme bromoperoxidase discovered in an alga (Vilter 1983) and a non-heme chloroperoxidase discovered in a fungus (Geigert *et al.* 1985). Clearly, although heme is normally a component of haloperoxidases, it is not an absolute requirement. For those who have studied peroxidases, this is not a totally unexpected finding, since over ten years ago, glutathione peroxidase (EC 1.11.1.9), a non-halogenating peroxidase, was also shown to be heme-free.

Although iron is the usual metal found in heme, two haloperoxidases have been found to contain significant levels of other metals. Thomas (1968), while running electron paramagnetic resonance studies on the chloroperoxidase from *C. fumago*, noted a complicated spectrum having splittings suggesting the presence of manganese (Mn^{2+}). He confirmed this observation by both emission spectroscopy and atomic absorption, showing that chloroperoxidase had manganese to enzyme ratios varying from 0.3 to 1.0. More recently, we have used X-ray fluorescence to confirm his results, and to show the absence of other transition metals (unpublished results). Does manganese play an active role in catalysis? This is still a unanswered question. However, it may be more than coincidental that the only haloperoxidase containing manganese is also the most versatile of these catalysts. That is, chloroperoxidase not only carries out the full range of halide oxidations, but also oxidizes primary alcohols to aldehydes, dismutates chlorine dioxide to chlorate, and converts alkenes directly to epoxides. De Boer and co-workers (1985), using electron paramagnetic resonance, showed that the algal bromoperoxidase of *Ascophyllum nodosum* contained vanadium (V^{5+}) as a prosthetic group, ligated to the enzyme in a 1:1 complex. This bromoperoxidase is reported to be markedly thermostable, but since it has not been compared side-by-side with other haloperoxidases, it is not clear if vanadium imparts any inherent benefit.

Horseradish peroxidase contains a significant level of a divalent alkali metal: two moles of calcium (Ca^{2+}) per mole of enzyme. The pyrrolidone carboxyl residue in the amino acid sequence of this enzyme may be responsible for the binding since proteins containing this residue, such as Factor VII and prothrombin, are known to complex Ca^{2+}. Therefore, this alkali metal most likely is not attached to the heme component but to the protein component. Although studies have shown that at least one calcium ion is necessary for the enzyme to maintain proper molecular conformation (Ogawa *et al.* 1979), it is not known whether calcium plays any role in catalysis, be it halogenation, peroxidation, or chlorite (ClO_2^-) utilization.

Haloperoxidases undergo the following reversible chemical reaction:

$$\text{native enzyme} \rightleftarrows \underset{\text{(prosthetic group)}}{\text{heme}} + \underset{\text{(apoenzyme)}}{\text{glycoprotein}}$$

The separation of the heme from the glycoprotein occurs readily at extremes of pH. However, in alkaline solution, the splitting is accompanied by irreversible denaturation of the glycoprotein, so reconstitution of the active enzyme is not possible. In acidic solution, with proper precautions, it is possible to carry out a reversible splitting of the enzyme. The general procedure is to slowly add the haloperoxidase (approximately 2 ml of a 1 mg protein/ml solution) into a mixture of acetone and HCl (400 ml of acetone containing 1.2 ml of concentrated HCl) in a dry-ice acetone bath. Using contrifugation at 4°C, the white glycoprotein precipitate is separated from the heme-containing supernatant. The precipitate is washed several times with fresh acetone and HCl solution and then dried. The heme supernatant is concentrated to a few yellow drops by vacuum aspiration of the acetone. The yellow drops (essentially heme plus concentrated HCl) are neutralized with NaOH. Both the white precipitate and yellow drops are brought up to 1 ml volume in phosphate buffer pH 6. Reconstitution of the original enzyme is accomplished by simply mixing equal volumes of the glycoprotein and heme solutions.

By substituting a synthetic heme for the original heme, synthetic enzymes ca be prepared:

$$\text{synthetic haloperoxidase} \longleftarrow \text{synthetic heme} + \text{glycoprotein}$$

This technique has been used for over 40 years, yielding many synthetic haloperoxidases containing metal- and porphyrin-substitutions (Figs. 4.2 and 4.3). The primary reason for preparing these enzymes has been to probe the nature of the active site; however, in a few cases the effect on peroxidase activity has also been examined. In those few cases, reduced peroxidase activity was always observed. Two limitations are noted from these studies. First, only horseradish peroxidase has been studied, although the technique should work on all heme-containing peroxidases. Second, only peroxidase activity has been measured and then only in a few cases, when reconstruction might have had a major effect on halogenation activity. In studies with other enzyme types, metal-substitutions have resulted in major changes in both specific activity and reaction specificity (Neidleman & Geigert 1984), so one might expect a similar effect with haloperoxidases. For example, by metal-substitution could horseradish peroxidase (which only iodinates) be converted into a haloperoxidase that oxidizes bromide ions? By metal-substitution could a bromoperoxidase be converted into a chloroperoxidase?

IIA	IIIA	IVA	VA	VIA	VIIA	VIIIA			IB	IIB
Mg										
			V		**Mn**	**Fe** (native)	**Co**	**Ni**	**Cu**	**Zn**
						Ru				

References:

Gjessing & Sumner (1942) (Mn, Co, Ni, Cu)
Gupta *et al.* (1980) (Mn)
Morishima *et al.* (1983) (Ru)
Tamura *et al.* (1977) (V)
Wang *et al.* (1977) (Co)
Yamazaki *et al.* (1982) (Mg, Zn)
Yonetani & Asakura (1969) (Mn)

Fig. 4.2 – Metal-substituted haloperoxidases that have been prepared (filled-in square indicates metal's position in periodic table).

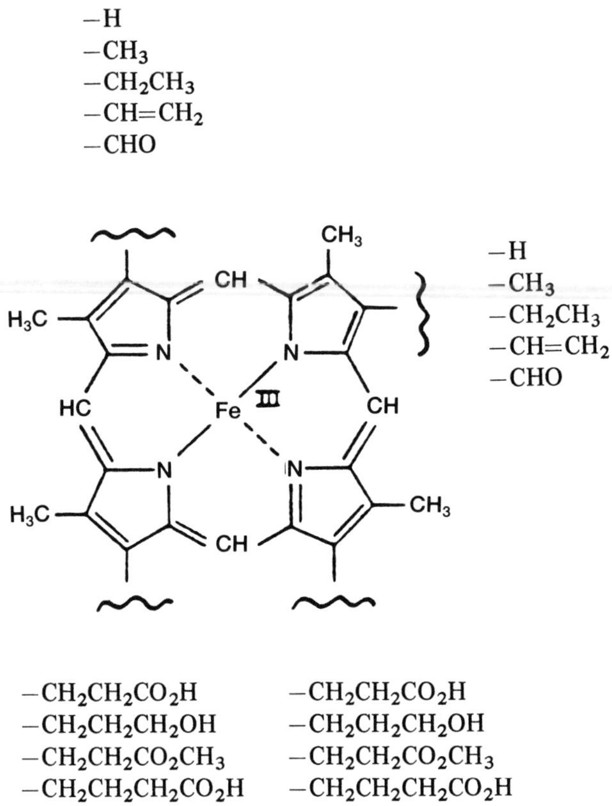

References: Araiso & Dunford (1981), DiNello & Dolphin (1981), Iizuka *et al.* (1982)

Fig. 4.3 – Some porphyrin-substituted haloperoxidases that have been prepared.

A different approach to metal substitution has been applied to the marine bromoperoxidase from *Ascophyllum nodosum.* The metal, vanadium, bound by the native enzyme was removed by dialysis against citrate-phosphate buffer with EDTA, resulting in complete inactivation of iodide oxidation activity. Then, various metal ions in pH 9 buffer were added to the metal-free enzyme. As shown in Table 4.2, full iodide oxidation activity was restored only when vanadium (V^{5+}) was added back (Vilter 1984).

Table 4.2 – Effect of metal-substitution on the bromoperoxidase of *Ascophyllum nodosum*.

Metal ions added to dialyzed enzyme	Reactivation of iodide oxidation activity (%)
V^{5+}	100
Co^{2+}	3
Ni^{2+}	1
Fe^{3+}	0
Fe^{2+}	0
Mn^{2+}	0
Cu^{2+}	0
Zn^{2+}	0
none	0

Reference: Vilter (1984).

The second component of haloperoxidases, the glycoprotein, plays a role both in determining the reactivity of the enzyme and in stabilizing the reactive intermediates. Some of the physical and chemical properties of this glycoprotein are summarized in Tables 4.3 through 4.6. Over the years, various apparent molecular weights have been reported for each of the haloperoxidases. For example, myeloperoxidase has been reported to have apparent molecular weights ranging all the way from 120 000 to 160 000 daltons. Sometimes the discrepancies can be accounted for by the method of measurement used; for example, Rawitch and co-workers (1979) and Schultz (1980) have shown that by gel filtration lower molecular weights are consistently assigned to haloperoxidases than by velocity sedimentation; and Nakagawa and co-workers (1985) have shown that the proteolytic digestion normally used to free membrane-bound thyroid peroxidase also cleaves off a portion of the native enzyme. The values listed in Table 4.3 are the more current ones; hopefully, the most reliable ones. The table shows that there is no direct correlation between molecular weight and halide ion specificity. For example, although chloroperoxidase, which oxidizes I^-, Br^-, and Cl^-, and horseradish peroxidase, which oxidises only I^-, are vastly different in their halide ion specificity, they have nearly identical molecular weights. The large size of some of the haloperoxidases may not be even necessary for catalytic activity. As an example, myeloperoxidase has been reductively cleaved in half, yielding hemi-myeloperoxidase (apparent molecular weight of 78 000 daltons). This half-enzyme exhibited the full peroxidase activity of the intact enzyme (Andews & Krinsky 1981), and more importantly, the half-enzyme also retained all of the chlorination activity of the intact enzyme (Andrews *et al.* 1984). A hemi-myeloperoxidase has been isolated directly from leukemia cells, but it is not known if it is identical to that produced by chemical reduction of myeloperoxidase (Yamada 1982).

Table 4.3 – Apparent molecular weight and subunits of haloperoxidase glycoproteins.

Enzyme	Apparent molecular weight	Subunits (4)	Reference
bromoperoxidase (*P. aureofaciens*)	158 000 (1)	77 000	van Pee & Lingens (1985)
myeloperoxidase	146 000 (2)	57 000 39 000 11 000	Merrill (1980) Olsen & Little (1984)
bromoperoxidase (*P. capitatus*)	97 600 (1)	55 000	Manthey & Hager (1981)
bromoperoxidase (*A. nodosum*)	95 000 (2)	–	Vilter (1983)
thyroid peroxidase	93 000 (1)	60 000 24 000	Rawitch *et al.* (1979)
lactoperoxidase	85 000 (1)	–	Dumontet & Rousset (1981)
eosinophil peroxidase	71 000 (1)	58 000 14 000	Bolscher *et al.* (1984a)
ovoperoxidase	70 000 (3)	–	Deits *et al.* (1984)
chloroperoxidase	42 000 (1)	–	Morris & Hager (1966b)
horseradish peroxidase	42 000 (1)	–	Aibara *et al.* (1982)

Method of measurement:

(1) Velocity sedimentation.

(2) Gel filtration.

(3) SDS-polyacrylamide gel electrophoresis (SDS-PAGE), nonreducing condition.

(4) SDS-polyacrylamide gel electrophoresis, reducing condition.

Table 4.4 – Amino acid analyses of haloperoxidases.

Amino acid residue	Symbol	Residues per mole of enzyme							
		Chloro-peroxidase[1]	Myelo-peroxidase[2]	Bromo-peroxidase[3]	Ovoper-oxidase[4]	Thyroid peroxidase[5]	Horseradish peroxidase[6]	Lacto-peroxidase[7]	Eosinophil peroxidase[8]
Aspartic acid*	Asp	39	132	40	68	59	45	71	80
Threonine	Thr	17	56	25	22	23	23	28	25
Serine	Ser	33	54	21	45	19	20	30	39
Glutamic acid†	Glu	26	136	33	54	67	21	60	38
Proline	Pro	23	84	2	42	69	18	42	38
Glycine	Gly	14	76	30	43	61	17	41	41
Alanine	Ala	24	72	30	40	74	23	40	50
Cysteine	Cys	2	24	16	12	13	8	16	13
Valine	Val	12	48	29	38	38	15	29	19
Methionine	Met	2	24	2	8	6	4	12	8
Isoleucine	Ile	9	32	20	31	18	12	28	24
Leucine	Leu	20	60	31	50	67	32	68	71
Tyrosine	Tyr	10	24	5	14	12	5	15	13
Tryptophan	Trp	ND	12	ND	15	13	1	16	ND
Phenylalanine	Phe	13	48	24	40	32	19	31	31
Histidine	His	7	12	3	17	20	3	14	9
Lysine	Lys	4	24	3	13	12	6	33	15
Arginine	Arg	6	96	16	48	51	20	39	81

References:
1. Morris & Hager (1966) (*C. fumago*).
2. Matheson *et al.* (1981).
3. Manthey & Hager (1981) (*P. capitatus*)
4. Deits *et al.* (1984).
5. Rawitch *et al.* (1979).
6. Aibara *et al.* (1982).
7. Rombauts *et al.* (1967).
8. Carlson *et al.* (1985).

ND = not determined.
* includes asparagine.
† includes glutamine.

Table 4.5 – Carbohydrate content of haloperoxidases.

Enzyme	Carbo-hydrate content (wt/wt)	Predominant sugars	Reference
chloroperoxidase (*C. fumago*)	13–17%	arabinose	Morris & Hager (1966b)
		glucosamine	Hashimoto & Pickard (1984)
horseradish peroxidase	14–17%	glucosamine	Aibara *et al.* (1982)
thyroid peroxidase	10%	mannose N-acetylglucosamine	Rawitch *et al.* (1979)
lactoperoxidase	5–10%	glucosamine	Rombauts *et al.* (1967)
		galactosamine	Carlstrom (1969)
ovoperoxidase	4%	mannose N-acetylglucosamine	Deits *et al.* (1984)
myeloperoxidase	3%	mannose, glucose N-acetylglucosamine	Bakkenist *et al.* (1978)
bromoperoxidase (*P. capitatus*)	0%	–	Manthey & Hager (1981)

Table 4.6 – Haloperoxidase isoenzymes.

Enzyme	Number of isoenzymes	Reference
horseradish peroxidase	42	Hoyle (1977)
lactoperoxidase	6–10	Carlstrom (1969)
chloroperoxidase (*C. fumago*)	3–6	Hashimoto & Pickard (1984)
myeloperoxidase	3	Bakkenist *et al.* (1978)
ovoperoxidase	2	Deits *et al.* (1984)

Several of the haloperoxidases contain subunits (i.e., multiple peptide chains). Bromoperoxidases from *P. capitatus* and *P. aureofaciens* are dimers having two peptide chains of equal size, while thyroid peroxidase and eosinophil peroxidase are dimers having two peptide chains of unequal size. Judging from the literature, most workers agree that myeloperoxidase (MPO) is a tetramer, but the nature of the peptide chains is confusing. Matheson *et al.* (1981) reported that MPO contained only one subunit (M_r 59 000). However, both Andrews & Krinsky (1981) and Bakkenist and co-workers (1978) reported two subunits for MPO, but of different apparent molecular weights (M_r 14 000, and 58 000; M_r 63 000 and 81 000, respectively). On the other hand, Olsen & Little (1983) reported that MPO contained three subunits (M_r 16 000, 39 000, and 57 000). Two recent studies appear to have resolved these discrepancies. Pember *et al.* (1983) were able to resolve human MPO into three forms, each of which contained different subunit patterns. Form I contained M_r 12 000 and 63 000 subunits, while forms II and III each contained M_r 12 000 and 57 500 subunits. This finding indicates that the different subunit patterns may be arising from MPO heterogeneity. However, this alone cannot explain all of the observed subunit patterns noted by the other workers. Olsen & Little (1984) have recently reported that the subunit pattern of MPO was a function of the way the enzyme was treated in the SDS-PAGE analysis. Reduction before heat treatment in SDS led to a two-subunit pattern (M_r 11 000 and 57 000), while reduction during or after heat treatment in SDS led to a three-subunit pattern (M_r 11 000, 39 000, and 57 000). However, heat treatment in SDS without any reduction yielded a single subunit (M_r 39 000). This study clearly shows the impact that sample treatment can have on the subunit pattern observed, and serves as a reminder of the pitfalls that can arise when comparing results between different laboratories.

Protein chemists, using the techniques of amino acid analysis, N-terminal and C-terminal amino acid sequencing, and proteolytic peptide mapping, have provided some insights into structural characteristics of the peroxidase proteins. Amino acid analyses of several haloperoxidases are presented in Table 4.4. Since these measurements were made in different laboratories at different times using various hydrolysis procedures, the potential for laboratory-to-laboratory variability dictates caution in any detailed comparison of the enzymes. However, the most striking general characteristic of these analyses is the predominance of acidic amino acids (aspartic acid and glutamic acid) over the basic residues (arginine and lysine), a ratio of greater than 2 for all of the reported haloperoxidases, except eosinophil peroxidase.

The complete amino acid sequence has been reported for horseradish peroxidase. Digestion with trypsin yielded 24 tryptic peptides, accounting for 203 of the approximate 300 amino acid residues of the enzyme (Welinder *et al.* 1972). Digestion with thermolysin yielded 120 thermolytic peptides, accounting for all but five of the amino acid residues of the enzyme (Welinder & Smillie 1972). With the necessary overlapping peptides, the complete amino acid sequence was pieced together (Fig. 4.4). Horseradish peroxidase is the only haloperoxidase so far studied in such detail. For lactoperoxidase, the N-terminal amino acid is reported to be leucine (Sievers 1981), with the tertiary structure maintained by eight disulfide bridges (Ohlsson & Paul 1983). For myeloperoxidase, the N-terminal amino acid on the heavy chain subunit is reported to be valine (Olsen & Little 1984), while that on the light chain subunit is reported to be lysine (Atkin *et al.* 1982).

Glu-Leu-Thr-Pro-Thr-Phe-Tyr-Asp-Asn-Ser-Cys-Pro-Asn-Val-Ser-
Asn-Ile-Val-Arg-Asp-Thr-Ile-Val-Asn-Glu-Leu-Arg-Ser-Asp-Pro-
Arg-Ile-Ala-Ala-Ser-Ile-Leu-Arg-Leu-His-Phe-His-Asp-Cys-Phe-
Val-Asn-Gly-Cys-Asp-Ala-Ser-Ile-Leu-Leu-Asp-Asn-Thr-Thr-Ser-
Phe-Arg-Thr-Glu-Lys-Asp-Ala-Phe-Gly-Asn-Ala-Asn-Ser-Ala-Arg-
Gly-Phe-Pro-Val-Ile-Asp-Arg-Met-Lys-Ala-Ala-Val-Glu-Ser-Ala-
Cys-Pro-Arg-Thr-Val-Ser-Cys-Ala-Asp-Leu-Leu-Thr-Ile-Ala-Ala-
Gln-Gln-Ser-Val-Thr-Leu-Ala-Gly-Gly-Pro-Ser-Trp-Arg-Val-Pro-
Leu-Gly-Arg-Arg-Asp-Ser-Leu-Gln-Ala-Phe-Leu-Asp-Leu-Ala-
Asn-Ala-Asn-Leu-Pro-Ala-Pro-Phe-Phe-Thr-Leu-Pro-Gln-Leu-Lys-
Asp-Ser-Phe-Arg-Asn-Val-Gly-Leu-Asn-Arg-Ser-Ser-Asp-Leu-Val-
Ala-Leu-Ser-Gly-Gly-His-Thr-Phe-Gly-Lys-Asn-Gln-Cys-Arg-Phe-
Ile-Met-Asp-Arg-Leu-Tyr-Asn-Phe-Ser-Asn-Thr-Gly-Leu-Pro-Asp-
Pro-Thr-Leu-Asn-Thr-Thr-Tyr-Leu-Gln-Thr-Leu-Arg-Gly-Leu-Cys-
Pro-Leu-Asn-Gly-Asn-Leu-Ser-Ala-Leu-Val-Asp-Phe-Asp-Leu-Arg-
Thr-Pro-Thr-Ile-Phe-Asp-Asn-Lys-Tyr-Tyr-Val-Asn-Leu-Glu-Glu-
Gln-Lys-Gly-Leu-Ile-Gln-Ser-Asp-Gln-Glu-Leu-Phe-Ser-Ser-Pro-
Asn-Ala-Thr-Asp-Thr-Ile-Pro-Leu-Val-Arg-Ser-Phe-Ala-Asn-Ser-
Thr-Gln-Thr-Phe-Phe-Asn-Ala-Phe-Val-Glu-Ala-Met-Asp-Arg-Met-
Gly-Asn-Ile-Thr-Pro-Leu-Thr-Gly-Thr-Gln-Gly-Gln-Ile-Arg-Leu-
Asn-Cys-Arg-Val-Val-Asn-Ser-Asn-Ser

Disulfide bridges: 11–91, 44–49, 97–301, 177–209.

Glu = pyrrolidone carboxyl N-terminal residue.

Reference: Welinder (1976).

Fig. 4.4 – Amino acid sequence of horseradish peroxidase.

The content of carbohydrate in the glycoprotein component ranges from near 0% to 50% (Table 4.5). Little is known about the chemical nature of this carbohydrate, beyond the percentage of various sugar monomers. For horseradish peroxidase, given the complete amino acid sequence, the eight sites of carbohydrate attachement to the protein have been assigned. These eight sites involve asparagine (Asn) at positions 13, 57, 158, 186, 198, 214, 255, and 268 (Welinder 1976).

Using analytical techniques such as isoelectric focusing, numerous isoenzymes have been found in the haloperoxidases (Table 4.6). In general, the isoenzymes of a given haloperoxidase have similar apparent molecular weights and catalytic rate constants. Differences in carbohydrate content, amino acid composition, and/or amino acid sequence seem to account for the existence of isoenzymes. In a few studies, significant differences in substrate specificities among isoenzymes have been noted. For example, it has been reported that the rate of reaction with a variety of substrates (i.e., ascorbic acid, uric acid, guaiacol, and coniferyl alcohol) varies among the isoenzymes of horseradish peroxidase (Chmielnicka *et al.* 1971). Is it possible that among the isoenzymes of a given haloperoxidase, major differences in halide oxidation specificity might be found? If so, then environmental control could lead to the production of specific isoenzymes. For example, in the terrestrial blue-green alga, *Phormidium ambiguisom*, the production of specific isoperoxidases was shown to be a function of the salt concentration in the growth medium (Anantani & Vaidya 1983).

CATALYTIC EXPRESSION

Catalysts, whether chemical or enzymatic, offer tremendous rate accelerations for chemical reactions compared to noncatalyzed rates. A useful term to describe enzymatic activity is the turnover number, more properly the catalytic center activity, expressed as μmoles of substrate reacted per minute per μmole of enzyme active site. With haloperoxidases, turnover numbers up to 10^5 min^{-1} are reported (Tables 4.7, 4.8). The observed multi-log range of acceleration is a function of each unique haloperoxidase reacting with the various substrates.

Table 4.7 – Turnover numbers for haloperoxidase-catalyzed iodination (μmoles product formed/min/μmole enzyme).

Enzyme	Turnover number	Reference
horseradish peroxidase	3.2×10^5	Bjorksten (1968)
lactoperoxidase	1.4×10^5	Bayse & Morrison (1971)
myeloperoxidase	1.4×10^5	Pember *et al.* (1983)
ovoperoxidase	1.4×10^5	Deits *et al.* (1984)
chloroperoxidase (*C. fumago*)	1.0×10^5	Hager *et al.* (1966)

The reaction:

$$3I^- + H_2O_2 + 2H^+ \xrightarrow{\text{enzyme}} I_3^- + 2H_2O$$

Measure absorbance at 350 nm due to appearance of I_3^-.

Table 4.8 – Turnover numbers for haloperoxidase-catalyzed chlorination and bromination (μmoles substrate reacted/min/μmole enzyme).

Enzyme	Turnover number	Reference
chloroperoxidase	$1.3 \times 10^5 (Br^-)$	Hager *et al.* (1966)
(*C. fumago*)	$6.7 \times 10^4 (Cl^-)$	
bromoperoxidase	$1.1 \times 10^5 (Br^-)$	Manthey & Hager (1981)
(*P. capitatus*)		
myeloperoxidase	$4.1 \times 10^3 (Cl^-)$	Wever *et al.* (1981)
eosinophil peroxidase	$1.4 \times 10^3 (Cl^-)$	Wever *et al.* (1981)
bromoperoxidase	$1.2 \times 10^2 (Br^-)$	van Pee & Lingens (1985)
(*P. aureofaciens*)		
ovoperoxidase	$4.4 \times 10^1 (Br^-)$	Deits *et al.* (1984)

The reaction:

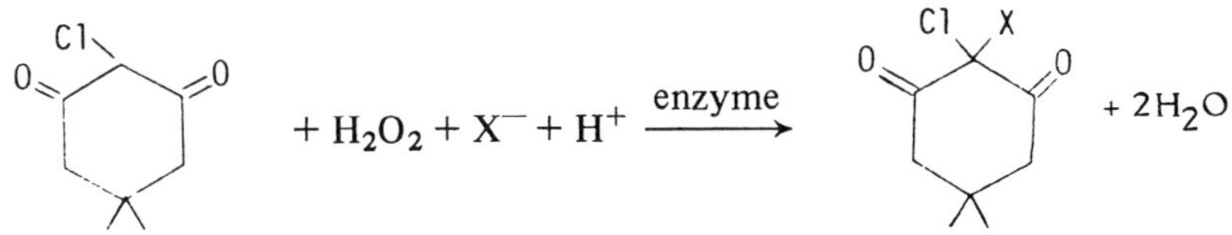

MCD

(where X = Cl or Br)

Measure disappearance of absorbance at 292 nm.

Normally, the pH at which the turnover number is determined is the optimal pH for the reaction. Most enzymes have a single optimal pH. With haloperoxidases, the situation is different since each of these enzymes has a range of optimal pH values rather than a single value. Two of the reactants in the haloperoxidase reaction, the halide ion concentration and the H_2O_2 concentration, dramatically affect the measured optimal pH value. For a given enzyme at a fixed halide ion concentration, the pH optimum shifts to lower pH values with increasing concentrations of H_2O_2; while at a fixed H_2O_2 concentration, the pH optimum shifts to higher pH values with increasing concentrations of halide ion. This interrelationship can be expressed by the following equation:

$$\text{optimal pH} = \beta + \log\left[\frac{[\text{halide ion}]}{[H_2O_2]}\right]$$

where β is the measured optimal pH when [halide ion] = $[H_2O_2]$. Thus, it is possible to vary the pH optimum one entire pH unit by simply changing the ratio of these two reactants ten-fold (Table 4.9). This shift has to be considered when running certain haloperoxidase reactions, for example, the conversion of alkenes to dihalides, as mentioned in Chapter 3. Since the reaction can only occur in high halide ion concentrations, this reaction must be run at a pH at least 1 pH unit higher than the other reactions in order to utilize the maximal acceleration offered by the enzyme. The effect of H_2O_2 and halide ion concentrations on the optimal pH is a common characteristic of the haloperoxidases. The phenomenon has been demonstrated to date on chloroperoxidase, myeloperoxidase, eosinophil peroxidase, and lactoperoxidase (Wever *et al.* 1980, Andrews & Krinsky 1982, Dunford *et al.* 1982, Lambeir & Dunford 1983).

Table 4.9 – Effect of the concentrations of H_2O_2 and Br^- on the pH optimum for the myeloperoxidase-catalyzed bromination of monochloro-dimedone.

$[Br^-] : [H_2O_2]$	Postion of the pH optimum
3	3.7
10	4.2
30	4.6
100	5.1
300	5.4
1000	5.9

Reference: Bakkenist *et al.* (1980).

A simple model to explain these effects has been proposed (Andrews & Krinsky 1982). The central feature of the model has two binding sites for halide ion. One is a substrate binding site leading to hypohalous acid formation, and the other is an inhibitor binding site, which can hinder the binding of H_2O_2 at the substrate binding site. However, binding of halide ion at the inhibitor binding site requires prior protonation of that site. An example of how the model works follows:

> Increasing halide ion concentrations makes more halide ion available for binding at the inhibitor binding site. This increased binding inhibits the binding of H_2O_2 at the substrate binding site. To overcome the inhibition, the pH must be raised to alter the charge at the inhibition site, thus minimizing the binding of halide ion at that site. The maximal acceleration is not changed, just the pH position at which the maximal acceleration occurs.

Each haloperoxidase has a preferred pH optimum region for reaction. The chloroperoxidase from *C. fumago* has the lowest pH region, approximately pH 3–4.

Why such a low pH is needed for maximal velocity is not certain, but one hypothesis suggests that an amino acid side chain of the chloroperoxidase protein donates a carboxylate ligand to the sixth coordinate position of the heme iron, blocking the binding of halide ion at the active site (Hager *et al.* 1970). At low pH, the carboxylate ligand is protonated and exits the sixth coordinate position. Most of the haloperoxidases prefer the neutral pH region (pH 4.0–7.5). In this region head-to-head comparisons at the same H_2O_2 and halide ion concentrations have shown detectable pH preferences among the haloperoxidases. For example, the pH optimum of myeloperoxidase is about 1 pH unit higher than that of eosinophil peroxidase (Wever *et al.* 1980). To date, no haloperoxidase has been discovered that prefers an alkaline pH region for reaction (i.e., pH > 8).

For a haloperoxidase-catalyzed reaction to occur, three reactants must be present: organic substrate, halide ion, and hydroperoxide. In Chapter 3, the broad range of organic substrates for these enzymes was presented. In addition, a range of halide ions and hydroperoxides can also be used in the reaction. In terms of halide ion reactants, haloperoxidases can be grouped according to their ability to oxidize specific halide ions:

chloroperoxidases:	Cl^-, Br^-, and I^- oxidized
bromoperoxidases:	Br^- and I^- oxidized
iodoperoxidases:	I^- only oxidized

These three groupings are patterned on chemical first principles. The standard oxidation–reduction potentials for reduction of H_2O_2 and oxidation of halide ions are as follows:

Half reaction	*E^o, volts*
$2F^- \rightleftharpoons F_2 + 2e^-$	−3.06
$H_2O_2 + 2H^+ + 2e^- \rightleftharpoons 2H_2O$	+1.77
$2Cl^- \rightleftharpoons Cl_2 + 2e^-$	−1.36
$2Br^- \rightleftharpoons Br_2 + 2e^-$	−1.07
$2I^- \rightleftharpoons I_2 + 2e^-$	−0.54

Therefore, a haloperoxidase having an oxidation-potential sufficient to oxidize chloride ions would be able to also oxidize bromide and iodide ions; an enzyme having an oxidation-reduction potential sufficent to oxidize bromide ions would be able to oxidize iodide ions but not chloride ions, etc. Yet there are occasions when the expected halide ion oxidation is not observed. The probable explanation is not that the enzyme cannot oxidize the expected halide ion, but that it oxidizes it too rapidly. The resulting excess hypohalous acid, a strong oxidant, attacks the enzyme as a suitable substate, quickly deactivating it; thus, no apparent reaction is measured (Ramakrishnan *et al.* 1983). An additional feature of these oxidation–reduction potentials is the conclusion that haloperoxidases using H_2O_2 cannot oxidize fluoride ions, despite a speculation to the contrary (Vickery *et al.* 1979). It should be noted, however, that although fluoride ion cannot be oxidized, it still is a halide ion, and as such, it competes with other halide ions present for the substrate binding site and acts like the other halide ions in hindering H_2O_2 binding as mentioned previously, thus inhibiting the measured halogenation activity (Zgliczynski *et al.* 1983).

In addition to the halide series, haloperoxidases can react with the pseudo-halide, thiocyanate (SCN^-). In the oxidation–reduction potential scale, oxidation of SCN^- falls between Br^- and I^-; therefore, every chloroperoxidase and

Table 4.10 – Hydroperoxides utilized by haloperoxidases.

	Value of k_1 Rate Constant (M^{-1} sec^{-1})			
Hydroperoxide (ROOH)	Horseradish peroxidase[1,2]	Chloro-peroxidase[3]	Myelo-peroxidase[4]	Thyroid peroxidase[5]
hydrogen peroxide (R = H)	2.0×10^6	2.2×10^6	2.0×10^7	7.5×10^6
methyl hydrogen peroxide (R = Me)	1.7×10^5	5.7×10^5	–	
ethyl hydrogen peroxide (R = Et)	7.6×10^4	–	5.6×10^3	
isopropyl hydrogen peroxide (R = $CH(Me)_2$)	3.1×10^2	–		
peracetic acid ($R = \overset{O}{\overset{\|}{C}}CH_3$)	8.8×10^7	3.8×10^6		
chloroperoxobenzoic acid ($R = \overset{O}{\overset{\|}{C}}-\emptyset-Cl$)	1.2×10^8			

References:

1. Paul *et al.* (1979).
2. Davies *et al.* (1976).
3. Araiso *et al.* (1981) (*C. fumago*).
4. Harrison (1982).
5. Nakamura *et al.* (1983).

Reaction: $\underset{\text{enzyme}}{\text{native}} + ROOH \xrightarrow{k_1} \underset{\text{I}}{\text{compound}} + ROH$

bromoperoxidase can oxidize thiocyanate to hypothiocyanous acid (HOSCN), while certain iodoperoxidases (thyroid peroxidase) can and other (horseradish peroxidases) cannot (Morrison & Schonbaum 1976).

In terms of hydroperoxide reactants, hydrogen peroxide (ROOH, R=H), because of its high reactivity and ready availability, is the hydroperoxide of choice. But there are other hydroperoxides that can also be utilized by haloperoxidases (Table 4.10). As an exception, the bromoperoxidase of *P. capitatus* was reported to react only with H_2O_2 (Manthey & Hager 1981). In studies with alkyl hydroperoxides, the purity of the hydroperoxide can be a problem. A simple trick to ensure that the alkyl hydroperoxide is, in fact, being examined is to add catalase to consume any contaminating H_2O_2.

REACTION MECHANISMS

Five oxidation–reduction states are known for haloperoxidases:

1. *Native resting state.* Iron is in the ferric form, Fe(III).
2. *Compound I.* When the native enzyme reacts with a hydroperoxide (ROOH) an enzyme intermediate, 2-equivalents oxidized above the native resting state, is formed. This intermediate, called Compound I, appears to be a ferryl porphyrin π-cation radical, wherein one of the oxidation equivalents is stored as the ferryl iron, Fe(IV), while the other equivalent is stored as a porphyrin-centered π-cation radical. Compound I is unstable (half-life of minutes or less) and has a characteristic green color.
3. *Compound II.* One-equivalent reduction of Compound I leads to Compound II. This intermediate is quite stable (half-life of hours) and has a characteristic red color.
4. *Compound III.* When the native enzyme reacts with molecular oxygen (O_2), or more correctly superoxide anion radical (O_2^-), an enzyme intermediate, referred to as Compound III, is formed. This intermediate is also called the oxy-ferrous complex.
5. *Compound X.* When the native enzyme reacts with chlorite (ClO_2^-), an enzyme intermediate, which contains one chlorine atom/enzyme molecule, is formed. This intermediate called Compound X, is very stable (half-life of days).

Some researchers would add a sixth oxidation–reduction state, called Compound EOX. In the next section, the possible existence of this halogenation intermediate will be discussed.

Because haloperoxidase can pass through each of these various oxidation–reduction states, numerous reaction mechanisms are possible. A total of five enzymatic cycles have been reported to date. By far, the two best known reaction classes of these enzymes are the peroxidations and the halogenations. The peroxidase cycle resembles a modified form of ping-pong kinetics in which one molecule of oxidizing substrate (H_2O_2) and two molecules of reducing substrate (AH) are involved:

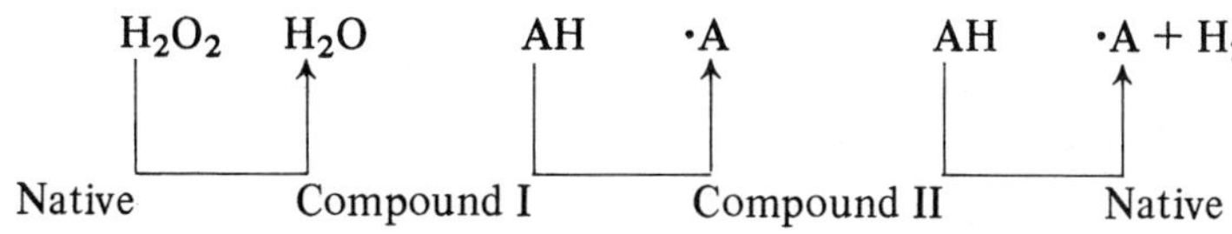

This leads to the overall reaction of:

$$H_2O_2 + 2AH \xrightarrow{\text{haloperoxidase}} A_2 + 2H_2O$$

The rate-determining step in peroxidations is the return of Compound II to the native enzyme. The halogenase cycle can be generally described as follows:

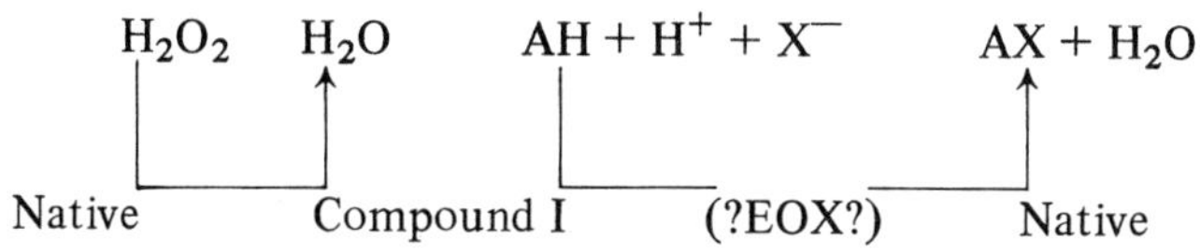

This leads to the overall reaction of:

$$H_2O_2 + AH + H^+ + X^- \xrightarrow{\text{haloperoxidase}} AX + 2H_2O$$

This halogenase cycle will be discussed in further detail in the next section.

Two of the haloperoxidases have been shown to contain a chlorite cycle: chloroperoxidase from *C. fumago* and horseradish peroxidase (Shahangian & Hager 1981, 1982). They utilize chlorite (ClO_2^-) in place of chloride ion and H_2O_2 to chlorinate the reducing substrate. The chlorite cycle is as follows:

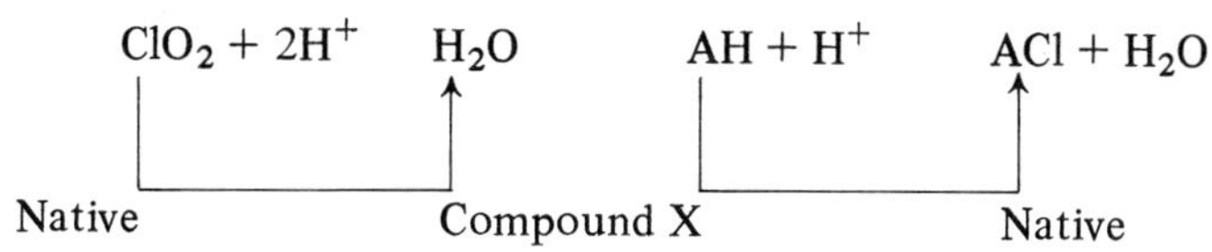

This leads to the overall reaction of:

$$ClO_2^- + 3H^+ + AH \xrightarrow[\text{or horseradish peroxidase}]{\text{chloroperoxidase}} ACl + 2H_2O$$

One intriguing point about this reaction is that it provides a pathway of chlorination for horseradish peroxidase, an enzyme that cannot directly oxidize chloride ion.

Less appreciated is the fact that haloperoxidases also have an oxidase cycle. The little known about this mechanism is that haloperoxidase will oxidize certain substrates (e.g., dihydroxyfumaric acid, indoleacetic acid) in the presence of molecular oxygen (O_2), and certain co-factors (e.g., Mn^{2+} , phenols) (Nichols 1962). Recent studies indicate that this oxidase cycle requires superoxide anion radical ($O_2^{\dot{-}}$) and proceeds through a reduced peroxidase $\rightleftharpoons$ Compound III shuttle (Smith *et al.* 1983). Removal of the heme from horseradish peroxidase leaves intact the ability of the remaining glycoprotein portion to still carry out this oxidase cycle while eliminating the peroxidase cycle (Siegel 1967).

Finally, one haloperoxidase, the chloroperoxidase from *C. fumago*, has a unique cycle, a true catalase cycle:

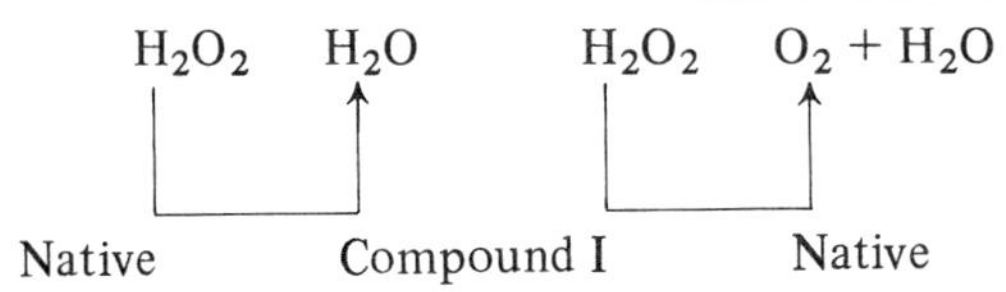

This leads to the overall reaction of:

$$2H_2O_2 \xrightarrow{\text{chloroperoxidase}} O_2 + H_2O$$

This is the only haloperoxidase reported to display such a reaction (Thomas *et al.* 1970). This novel reaction with chloroperoxidase should not be confused with the general halide-dependent pseudo-catalase side reaction common to all haloperoxidases (Carlsson 1983, Huwiler & Kohler 1984, Kanofsky 1984). In this latter case, the enzyme-formed halogenating intermediate, hypohalous acid (HOX), reacts with excess H_2O_2 to release O_2 as follows:

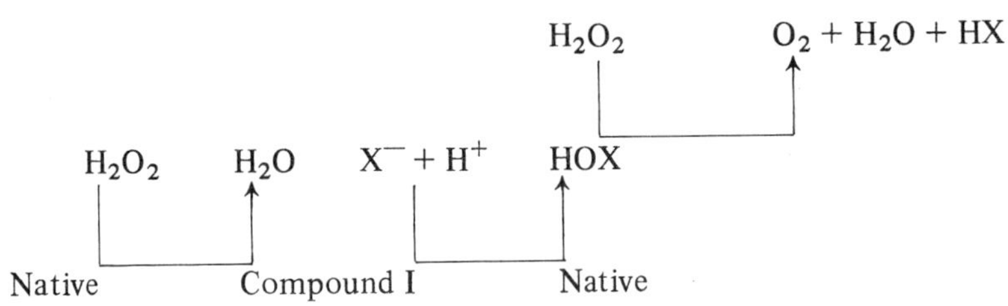

This leads to the same overall reaction as the catalase cycle, but by a halide-dependent pathway.

The complete picture of the enzymatic cycles of haloperoxidases is summarized in Fig. 4.5. This diversity in reaction mechanisms is a mixed blessing. On the one hand, the haloperoxidases have increased utility in that they can be used to carry out various reaction classes – peroxidation, halogenation, and oxidation. On the other hand, the haloperoxidase reaction has to be carefully controlled to insure that the desired mechanism is favored.

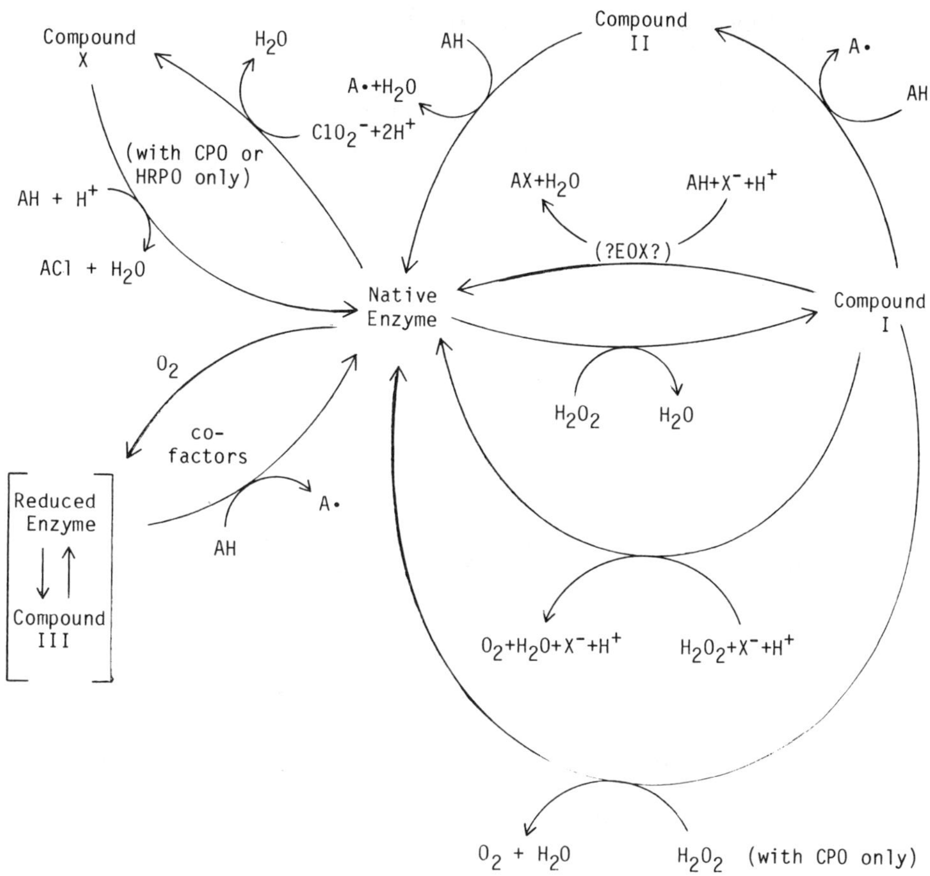

Fig. 4.5 – Reaction cycles of haloperoxidases.

THE HALOGENATING INTERMEDIATE

The two most plausible mechanisms for the halogenase cycle of haloperoxidases are presented in Fig. 4.6. Both pathways involve Compound I. If hypohalous acid (HOX) is the halogenating species as shown in mechanism A, then the reaction rate, substrate specificity, and product selectivity of the enzyme reaction should be identical with that of chemical halogenation processes. On the other hand, if an enzyme-bound intermediate (Compound EOX) is the halogenating species as shown in mechanism B, then the reaction rate, substrate specificity, and product selectivity of the enzyme reaction should be quite different from that of chemical halogenation processes. Considering the many studies that have been done, one would think that by now one of the mechanisms would have been eliminated from consideration, but such is not the case.

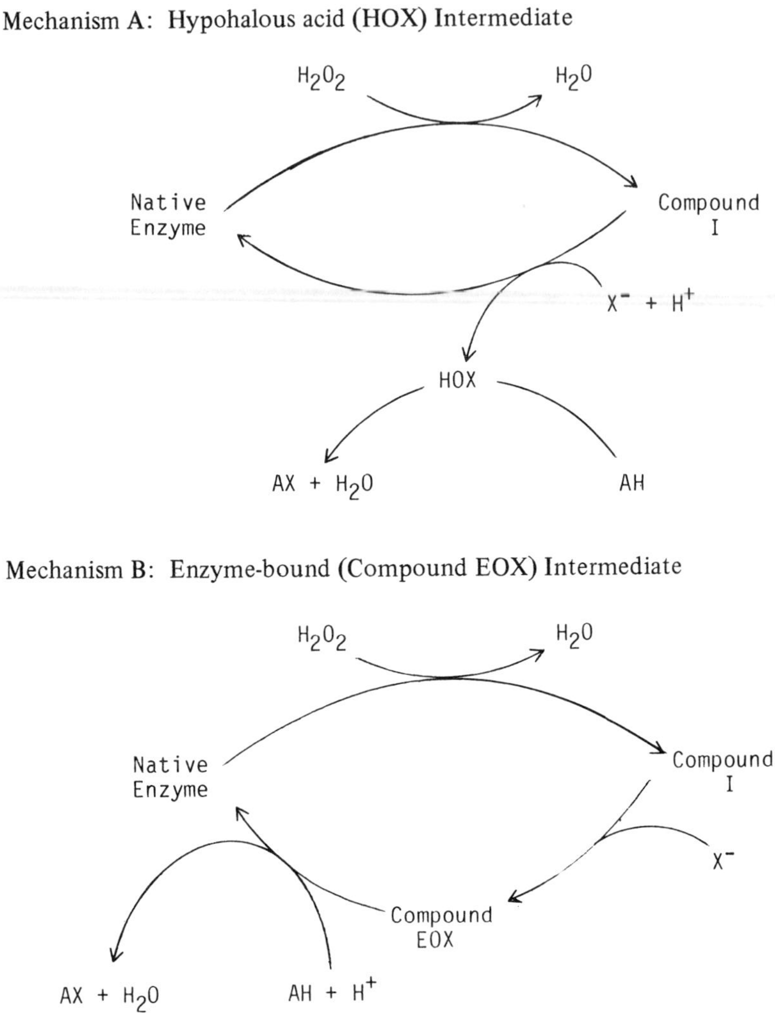

Fig. 4.6 – The two most plausible mechanisms for haloperoxidase-catalyzed halogenation.

For myeloperoxidase, mechanism A (i.e., HOX formation) is the generally accepted pathway for halogenation owing to the convincing mechanism studies carried out by Harrison & Schultz (1976). When enzyme, Cl^-, and H_2O_2 were mixed, the characteristic odor of bleach (i.e., HOCl) was noted. Substituting radioactive chloride ($^{36}Cl^-$) for Cl^- permitted measurement of the volatile radioactive end product, $^{36}Cl_2$. However, these results were only indicative of the reaction mechanism. To prove conclusively that mechanism B (i.e., enzyme-associated halogenation species) was not involved, the rate of HOCl formation by the enzyme had to be shown to be equal to the rate of chlorination of the substrate. To accomplish this, the produced HOCl was removed from the enzyme, since the enzyme itself is a substrate that can be chlorinated and inactivated. This was achieved by using myeloperoxidase immobilized on a Millipore membrane and pumping through a solution of substrate, H_2O_2 and Cl^-. Under set conditions (90 μg of immobilized myeloperoxidase, 200 μM H_2O_2, and 50 mM Cl^-), a turnover number of 100 min^{-1} was obtained both in the absence of substrate (i.e., HOCl was measured in the effluent) and in the presence of substrate (i.e., chlorinated taurine was measured in the effluent). Thus, the rate of HOCl formation was shown to be equal to the rate of substrate halogenation, and mechanism A was unequivocally shown for this haloperoxidase.

For chloroperoxidase from *C. fumago*, while there are points of general agreement, there are some major points of disagreement on the mechanism of halogenation. In the absence of a substrate, the enzyme is believed to catalyze the oxidation of halide ion to hypohalous acid. In the case of iodide, molecular iodine is believed to be the intermediate, even where an acceptor substrate is present. For this, there is general agreement. The disagreement involves the mechanism for the bromination and chlorination of substrates. Those who study substrate halogenation rates favor mechanism B, while those who study the nature of the halogenated products prefer mechanism A. Hager and co-workers have a dual argument for the involvement of an enzyme-bound halogenating intermediate, i.e., mechanism B (Libby *et al.* 1982). First, the rate of enzymatic oxidation of chloride to its respective molecular species (i.e., HOCl) was found to be very slow compared to the rate of chlorination of the substrate, 2-chlorodimedone. Clearly, this result is in opposition to that obtained in the myeloperoxidase study. However, the study with chloroperoxidase differed from that for myloperoxidase. Since HOCl can readily attack chloroperoxidase thus both affecting the rate of HOCl formation and consuming HOCl, and since HOCl is decomposed in reaction with H_2O_2, it is not possible to accurately quantitate the true amount of HOCl formed in the chloroperoxidase study. It would be most interesting to see how chloroperoxidase behaves in the immobilized myelopeoxidase flow system previously described, thus avoiding these problems. Second, in a competitive substrate study, it was observed that major differences in substrate specificity occurred between enzymatic and chemical chlorination (Table 4.11). Such a result is expected if an enzyme-associated halogenating intermediate is involved. However, much smaller differences were noted in the bromination comparisons. The difficulty in interpreting these substrate specificity results in deciding which chemical halogenation condition to use. Specificity of a chemical reaction can vary, depending upon the experimental conditions. For example, it has been shown that shifting from pH 4.0 and pH 7.0 in the chlorination of phenol with HOCl in aqueous solution at 20°C, results in a 100% shift in the *ortho*/*para* ratio for the products formed (Ogata *et al.* 1984). The concentration of the halogenating intermediate is another variable to consider. Was the mixing efficiency of the chemical halogenation reaction sufficient? If not, then locally concentrated zones of chemical reagent could obscure the true specificity. Mixing efficiently would not be a problem for the freely dispersed enzyme, since it would produce the halogenating intermediate slowly and continuously in a homogeneous solution.

Table 4.11 – Comparison of substrate specificity: chloroperoxidase versus chemical halogenating reagents.

	Substrate specificities*				
	Chlorination			Bromination	
Substrate pair	Chloro-peroxidase	HOCl	Cl_2	Chloro-peroxidase	Br_2
thiourea:MCD	>50:1	2:1	3.1	>50:1	>50:1
methionine:MCD	>50:1	1.5:1	–	4:1	1.5:1

* Relative preference of halogenating agent for each pair of substrates at equal concentrations.

Reference: Libby *et al.* (1982).

Reactions:

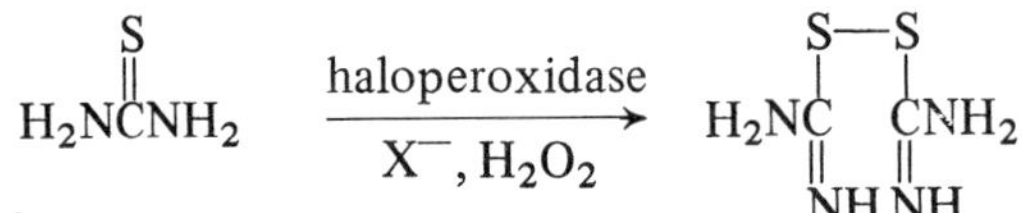

thiourea

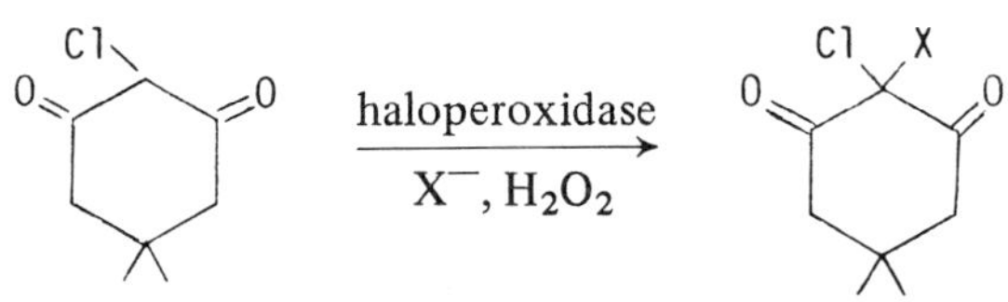

MCD

$$CH_3SCH_2CH_2CH(NH_2)CO_2H \xrightarrow[X^-,\ H_2O_2]{\text{haloperoxidase}} CH_3\overset{O\uparrow}{S}CH_2CH_2CH(NH_2)CO_2H$$

methionine

Lambeir & Dunford (1983), after studying the initial rate steady state kinetics for the chlorination of 2-chlorodimedone (MCD) by chloroperoxidase, also reached the same conclusion as Hager. In these experiments the concentrations of MCD, chloride ion, and peracetic acid were systematically varied under steady-state conditions, and the effect on the initial rates of reaction was observed. The initial rates increased with increasing concentration of both chloride ion and MCD. From these kinetic parameters, the model involving an enzyme-bound halogenating intermediate was favored. If haloperoxidases were normal enzymes, these results would definitely support the presence of an enzyme-bound halogenating intermediate (Compound EOX), especially in explanation of why MCD concentration affects the initial rates. However, because these enzymes have multiple pathways, the data could be explained in other ways. The change in initial rates upon varying chloride ion concentration could be explained by a shift in the pH optimum for the reaction due to the shift in the $[Cl^-]:[ROOH]$ ratio. In this study the peracetic acid concentration was held at 1 mM, while the chloride ion concentration varied from 0.005 to 0.06M, yielding approximately a ten-fold shift in the $[Cl^-]:[ROOH]$ ratio. Since the pH was held at 3.2 during this study, the optimum pH, hence the maximal reaction velocity, would shift from pH 2.8 up to pH 3.2 as the chloride ion was increased, resulting in an apparent increase in the initial rates. The change in initial rates upon changing the MCD concentration could be explained by a multiple substrate model. Initial rates were determined by monitoring the disappearance of MCD, so the reaction of HOCl with any other substrate would not be detected. Since enzyme and excess peracetic acid, along with MCD, are all suitable substrates for HOCl, increasing the MCD concentration would give the compound a competitive edge in the reaction, resulting in an apparent increase in initial rates. In addition, Lambier & Dunford found that the rate constant for the reaction of the presumptive Compound EOX with MCD was a factor of 200 times greater than the reaction of MCD with added HOCl. Why the uncatalyzed reaction with HOCl is so much slower is different to explain unless an enzyme-bound halogenating intermediate did, in fact, exist. On the other hand, Griffin & Ashley (1984) argue that the compaison was not correctly controlled. The rate of reaction of MCD with added HOCl was measured in the absence of Cl^-, which is not the proper control for the enzymatic reaction that contains a high concentration of Cl^-. In the presence of Cl^-, the steady-state concentration of Cl_2 in equilibrium with HOCl is greater, and since the rate of reaction of MCD with Cl_2 is much greater than with HOCl, this could eliminate any difference in observed rates of reaction between the chemical process and the enzymatic system.

In an opposing viewpoint, those who have studied product selectivity with chloroperoxidase favor mechanism A (i.e., HOX formation). As early as 1970, Kollonitsch and co-workers noted that this enzyme failed to form optically active chlorohydrins when reacted with propenylphosphonic acids. Subsequent enzymatic studies on other alkenes have confirmed this finding (Fig. 4.7). Neither the bromohydrins of propylene and styrene (unpublished results, Cetus) nor the bromohydrin of *trans*-cinnamic acid (personal communication, H. H. Freeman, Dow Chemical) were found to be optically active. Recently, Ramakrishnan and co-workers (1983), in an extensive study, documented the absence of stereoselectivity with chloroperoxidase in the reaction of additional substrates – two bicycloalkenes, methionine, and a β-diketone (Fig. 4.8). This lack of stereoselective product formation is consistent with a chemical halogenating intermediate like HOX. In addition to stereoselectivity, there is also the possibility of geometric-selectivity in the products formed. Substrates such as unsymmetrical alkenes and conjugated alkenes can form positional, halohydrin isomers. However, in every alkene that has been studied, the ratio of positional isomers formed by chloroperoxidase matched that formed by merely adding hypohalous acid to the

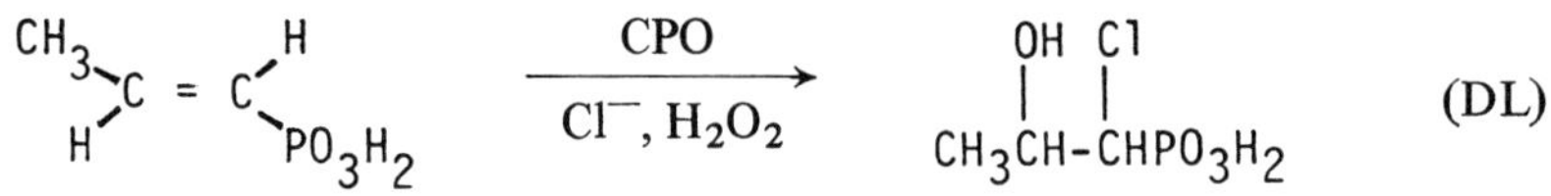

trans-propenylphosphonic acid

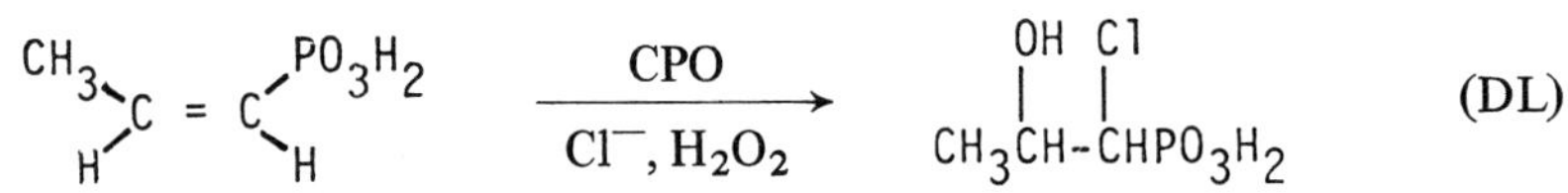

cis-propenylphosphonic acid

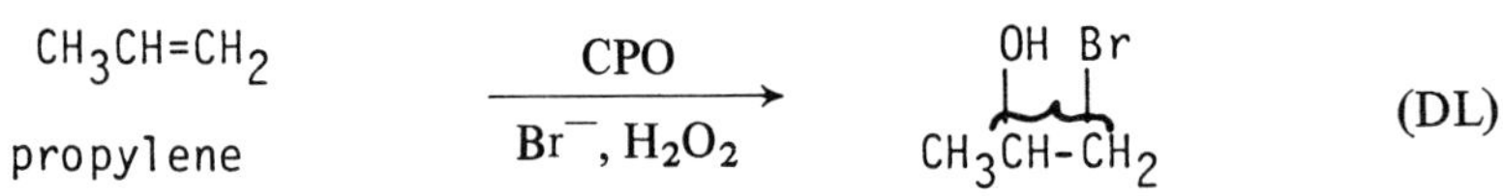

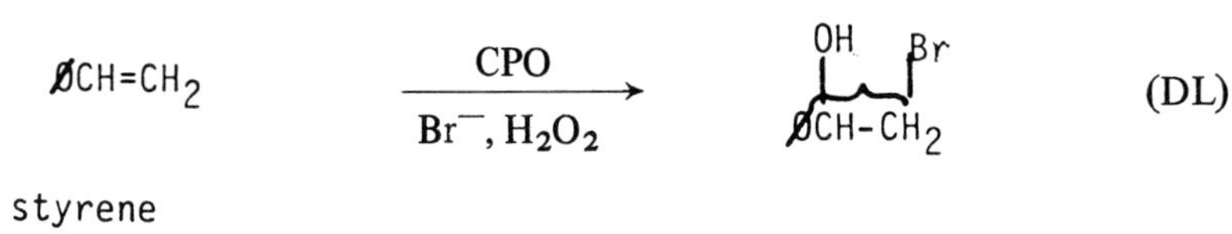

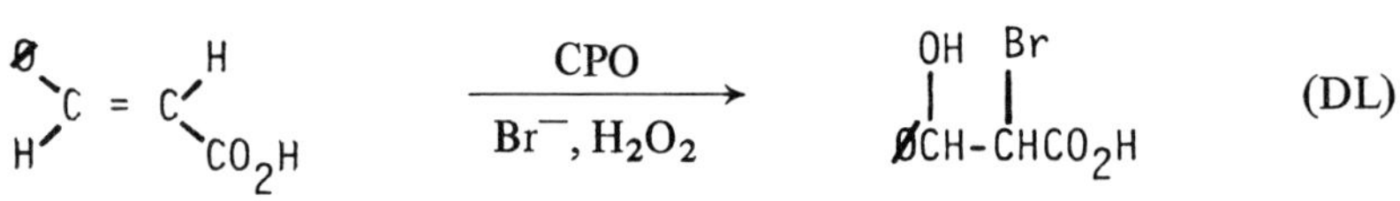

trans-cinnamic acid

Fig. 4.7 – Absence of stereoselective product formation with chloroperoxidase-catalyzed halogenation of alkenes.

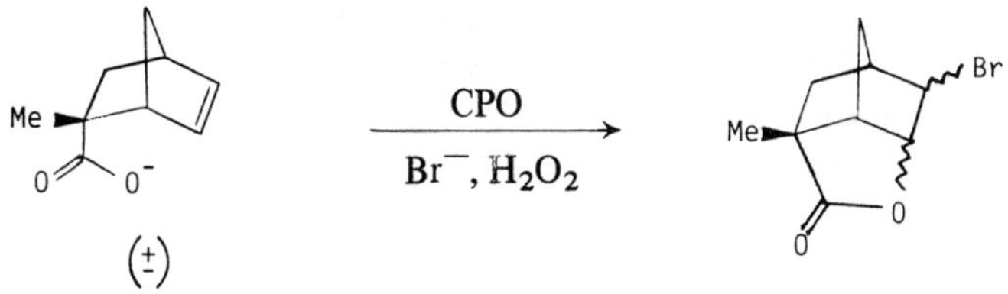

2-*exo*-methylbicyclo(2.2.1)hept-5-ene;
endo-carboxylic acid

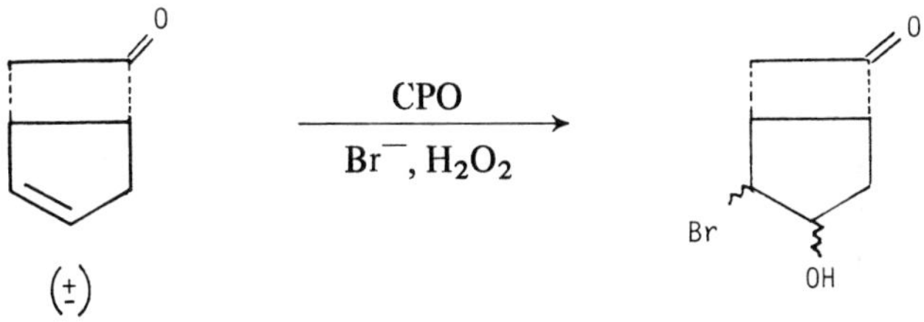

bicyclo(3.2.0)hept-2-en-6-one

$$CH_3SCH_2CH_2CH(NH_2)CO_2H \xrightarrow[Cl^-,\ H_2O_2]{CPO} CH_3\overset{\overset{O}{\uparrow}}{S}CH_2CH_2CH(NH_2)CO_2H$$

L-methionine

(50: 50 diastereomeric composition formed)

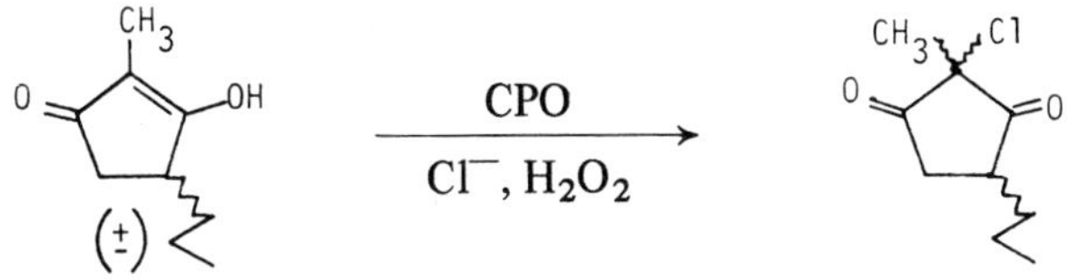

2-methyl-4-propylcyclopentane-1,3,-dione

Reference: Ramakrishnan *et al.* (1983).

Fig. 4.8 – Racemic product formation with an additional four substances in chloroperoxidase-catalyzed halogenation.

alkene (Geigert *et al.* 1983b, Lee *et al.* 1983a). Substrates such as phenol and its derivatives may also form positional, halogenated isomers. However, Brown & Hager (1967) showed that the reaction of chloroperoxidase with Cl^- and anisole yielded the same *para*:*ortho* ratio of monochlorinated anisole derivatives as that obtained by simply adding HOCl to anisole. In fact, it is interesting to note that Breslow (1982) has since reported that an "artificial enzyme' (HOCl plus cyclodextrin) shows a more typical enzyme-like selectivity in the chlorination of anisole than does chloroperoxidase itself. Although these product selectivity studies favor mechanism A, the results in themselves do not prove that the favored enzymatic pathway involves HOX reaction. This can only be done if the rate of HOX formation by the enzyme equals the rate of product formation, and such studies with those substrates and chloroperoxidase have not been done.

Unlike myeloperoxidase- and chloroperoxidase-catalyzed iodination, there is no consensus for the mechanism of iodination catalyzed either by horseradish peroxidase or by thyroid peroxidase. Davidson and co-workers (1978), Nakamura and co-workers (1983), as well as Magnusson and co-workers (1984), all argue for the presence of an enzyme-bound iodinium cation as the halogenating intermediate (i.e., mechanism B), while Dunford & Ralson (1983) argue for the formation of hypoiodous acid as the halogenating intermediate (i.e., mechanism A). Each conclusion was drawn from either studies of substrate reactivities or rates of iodide oxidation.

It is evident that the mechanism of halogenation by haloperoxidases remains unresolved. Although this controversy as to whether HOX or Compound EOX is the key halogenating intermediate will continue, it is safe to say that all products produced and *all* product ratios obtained (whether geometric or stereo) by haloperoxidase-catalyzed halogenations can be explained by hypohalous acid chemistry.

5

The role of halogenating enzymes and halometabolites in the marine environment

INTRODUCTION

The marine environment is the most prolific source of halogenated compounds. Some of these have been illustrated in Chapter 2. Table 5.1 indicates the progress being made in isolating and characterizing halometabolites from marine sources, as reported in various reviews. The questions to be considered in this discussion are: How are these marine halometabolites synthesized, and how do they function in the interest of the producing organism?

Table 5.1 – Increase in isolated halometabolites from marine sources.

Number of halometabolites reported	Reference
0	Petty (1961)
~50	Siuda & DeBernardis (1973)
140	Faulkner (1976)
~300	Thomson (1978)
~700	Hager (1982)

ASPECTS OF THE BIOSYNTHESIS OF MARINE HALOMETABOLITES

The marine halometabolites can be characterized as falling into five major chemical classes: hydrocarbons, phenolics, ketones, terpenes, and 15-carbon nonterpenes (Hager 1982). As might be anticipated from the reactions discussed in Chapter 3, most of these halometabolites could result from enzymatic modification of alkenes, alkynes, phenols, and β-ketoacids.

Among this diversity of marine halometabolites, there are a considerable number of heterogeneous halides, that is, compounds in which two different halides have been incorporated (see, for example, Chapter 2, Fig. 2.3). Although halogenating enzymes had been studied for over twenty years, it was not until recently (Neidleman & Geigert 1983a) that an enzymatic method to accomplish heterogeneous halogenation, in contrast to homogeneous halogenation, was demonstrated in the laboratory.

These factors, in combination, contributed to optimizing the enzymatic synthesis of various heterogeneous dihalide derivatives. They were:

1. The nature of the reaction substrate.
2. The use of a specific subclass of halogenating enzymes.
3. The ratio of the concentrations of the two, different halide ions.

When three closely related substrates, propylene, allyl chloride, and allyl alcohol were examined, the data in Table 5.2 were obtained (Neidleman & Geigert 1983a). The data illustrate that, depending upon whether the biosynthetic preference of the organism is towards bromochloro- or bromo-derivatives, either an 'activated' alkene (allyl alcohol) or an 'unactivated' alkene (propylene), respectively, would be optimal. The experimental enzymatic synthesis of heterogeneous dihalides was greatly facilitated by the discovery of allyl alcohol as a substrate.

The approximate concentrations of Cl^- and Br^- in seawater are 500 mM and 1 mM, respectively. Therefore, in the absence of selective concentration or exclusion of one of these halide ions, an enzyme is reacting in the presence of a $Cl^-:Br^-$ molar ratio of 500. Table 5.3 shows the results from studies comparing the capacities of a chloroperoxidase (CPO from *C. fumago*) and a bromoperoxidase (LPO from milk) to synthesis bromochloro-allyl alcohols at various $Cl^-:Br^-$ molar ratios (Neidleman & Geigert 1983a).

A number of comments may be made concerning these data. As the molar ratio of $Cl^-:Br^-$ approaches that of unaltered seawater:

1. The bromoperoxidase afforded a greater percent conversion to the bromochloro-derivatives than the chloroperoxidase.
2. The bromoperoxidase synthesized no chlorinated side-products.
3. The chloroperoxidase synthesized a substantial amount of chlorinated derivatives.

All of the halogenating enzymes from marine sources that have been acceptably characterized have been bromoperoxidases. This, in conjunction with a $Cl^-:Br^-$ molar ratio of 500, optimizes for the production of bromochloro-derivatives.

There is a potential disadvantage to a marine organism in synthesizing bromochloro-derivatives or bromo-derivatives at such a high $Cl^-:Br^-$ molar ratio. Since the available Br^- is so low, it is necessary to process a relatively high volume of seawater in order to produce a substantial amount of the desired

Table 5.2 – Effect of substrate on enzymatic synthesis of halogenated derivatives.

	% of total halogenated products	
	Heterogeneous	Homogeneous
Substrate ($RCH = CH_2$)	$R{-}CH(Cl){-}CH_2(Br)$	$R{-}CH(OH){-}CH_2(Br)$ + $R{-}CH(Br){-}CH_2(Br)$
propylene ($R = CH_3$)	21	79
allyl chloride ($R = CH_2Cl$)	39	61
allyl alcohol ($R = CH_2OH$)	86	14

Enzyme = Lactoperoxidase; $Cl^-:Br^-$ molar ratio = 100.

Table 5.3 – Effect of $Cl^-:Br^-$ ratio on the synthesis of halogenated derivatives of allyl alcohol by chloroperoxidase (CPO) and lactoperoxidase (LPO).

	$Cl^-:Br^-$ molar ratios					
	1		100		333	
Products	CPO	LPO	CPO	LPO	CPO	LPO
bromohydrins and dibromo- derivatives	32*	39	11	14	3	8
bromochloro- derivatives	59	61	82	86	73	92
chlorohydrins and dichloro- derivatives	9	0	7	0	24	0

* values expressed as percent of total products.

compounds. One answer is to decrease the $Cl^-:Br^-$ molar ratio by either selectively concentrating Br^- or selectively excluding Cl^-. It has been reported that in specialized vesicular cells of the *Bonnemaisoniacea*, bromide ion is increased 30-fold, iodide ion is increased 3-fold, while chloride ion is reduced 2-fold, as compared to the quantities in adjacent vegetative cells not specialized in this function (Wolk 1968). Under these conditions, the advantage of the bromoperoxidases, as compared to the chloroperoxidase, in producing bromo-chloro-derivatives would be minimized, but since no chlorinated side-products are formed, the percent of the total products that would contain bromine would still be optimized, and this may be in the interest of the organism. The utility of brominated intermediates in biosynthesis may be advantageous owing to the greater reactivity of bromine as compared to chlorine.

Another variable that should be included as a determinant of the chemical nature of halometabolites is the level of hydrogen peroxide (H_2O_2) available to the halogenating system. This is demonstrated by the data in Table 5.4 for the reaction of chloroperoxidase (CPO) with the alkyne, methyl acetylene. It can be seen that at a given level of enzyme, substrate, and halide ion, increasing H_2O_2 concentration increases the production of dihalogenated products as compared to monohalogenated products (Geigert *et al.* 1983d).

Table 5.4 – Effect of H_2O_2 concentration on the chlorination of methyl acetylene ($CH_3{-}C{\equiv}CH$).

	Ratio of products formed		
Initial H_2O_2 (mM)	CH_3COCH_2Cl :	$CH_3COCHCl_2$:	$ClCH_2COCH_2Cl$
3	10	8	2
30	10	23	3
150	10	91	22

Yet another point to be emphasized is that the concentrations and ratio of halide ions vary from seawater values in many other high-halide niches. For example, the Dead Sea contains Cl^- at 5700 mM and Br^- at 60 mM for a $Cl^-:Br^-$ molar ratio of 95 (Neidleman & Geigert 1983b). Is there an effect of these variations in $Cl^-:Br^-$ molar ratios on the design of the halogenating enzymes or in the distribution of classes of halometabolites as produced by populations of organisms in these various niches?

In so far as the mechanism for enzymatic formation of these heterogeneous dihalides is concerned, it has been postulated that peroxidative bromination initiates the formation of bromochlorometabolites (Faulkner 1976). A biosynthetic sequence has been proposed that consists of addition of the bromonium ion (Br^+) to an olefinic bond, followed by addition of the nucleophilic chloride ion (Cl^-) to the intermediate, as shown in Fig. 5.1. It should be mentioned that there are chlorinated marine halometabolites (for example, see Chapter 2, Fig. 2.3). Are these synthesized by undiscovered chloroperoxidases, selective debromination of bromochloro-derivatives, or by halide exchange reactions?

Like bromonium ion-induced biosynthesis, iodonium ion-induced biosynthesis can occur. Here, an enzyme that can only oxidize I^- (not Br^- or Cl^-), an iodoperoxidase, can passively incorporate Cl^- as well as Br^-, as shown in Fig. 5.2. In addition, halogen exchange with Cl^- from seawater is far easier on organic-bound iodine than on organic-bound bromine. Such a reaction sequence is postulated to occur in the formation of chloromethanes from iodomethanes in the marine environment (Theiler *et al.* 1978b).

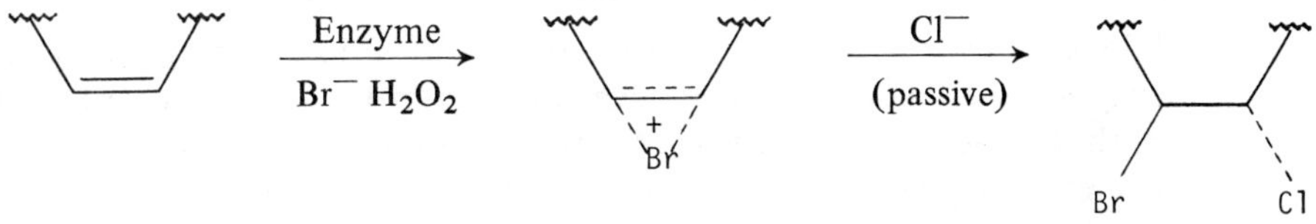

Reference: Faulkner (1976).

Fig. 5.1 – Proposed bromonium ion-induced enzymatic synthesis of bromochloro metabolites.

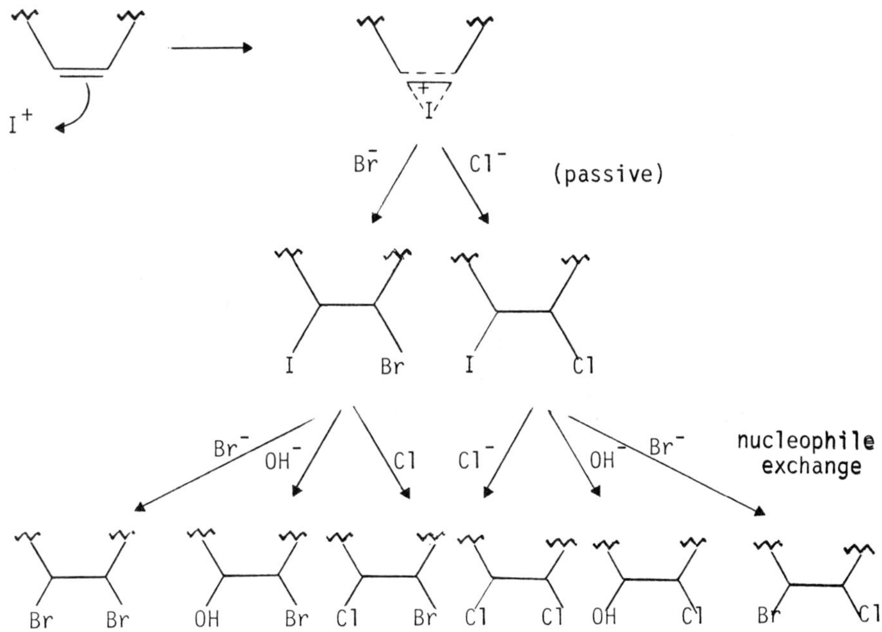

Reference: Geigert *et al.* (1984).

Fig. 5.2 – Proposed iodonium ion-induced biosynthesis of bromoiodo- and chloroiodo-metabolites, and their respective mono-, and heterogeneous and homogeneous dihalo-derivatives.

One wonders why marine organisms synthesize bromochloro-derivatives. Do these heterogeneous halides have some unique biological properties not possessed by their homogeneous counterparts? Is the distribution of halogenated products and bromoperoxidases in the marine environment a passive or an active response to the stresses of nature and the struggle for survival?

HALOGENATED INTERMEDIATES IN BIOSYNTHESIS

Why is it in the interest of marine organisms to invest energy in halogenating enzymes and halometabolites? There are two major reasons: 1) as key elements in the biochemical synthesis of certain metabolites, and 2) as a way of producing compounds with enhanced and desirable biological activity.

Fenical (1979, 1982) has speculated that several biosynthetic reactions, including the following, appear to involve brominated intermediates formed via bromonium ion (Br^+) insertion:

1. The synthesis of linear and branched -enoic acids by Favorsky rearrangements of brominated ketones (Fig. 5.3), as well as lactones, γ-pyrones, and octen-3-ones.
2. The cyclization of linear isoprenoids to yield a variety of terpenoids (Fig. 5.4).
3. The ring contraction of the diterpenoid concinndiol to a hydroperoxide derivative (Fig. 5.5).
4. The methyl migration in the terpene α-synderol to yield rearranged terpenoid derivatives (Fig. 5.6).

The latter two proposals are particularly interesting as examples of unbrominated metabolites arising through brominated intermediates. Neidleman (1975) stated:

> Microorganisms, it would seem, should have discovered the fact, as have organic chemists, that halogenated intermediates, on pathways to non-halogenated end products are useful devices. Little or no work to test this premise has been done. The halometabolites so far isolated and identified are mainly stable end products.

The work of Fenical and his co-workers is beginning to reveal the truth.

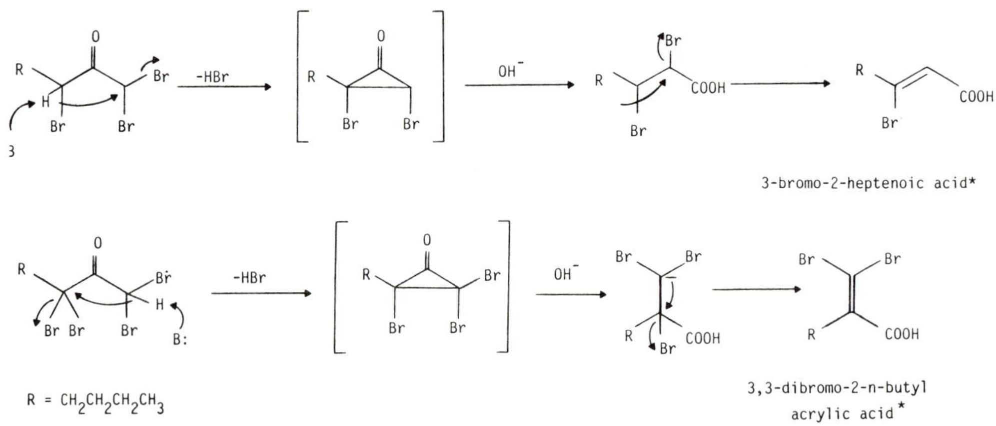

* Both compounds isolated from *Bonnemaisonia nootkana.*
Reference: Fenical (1979).

Fig. 5.3 – Synthesis of linear and branched-enoic acids by Favorsky rearrangements of brominated ketones.

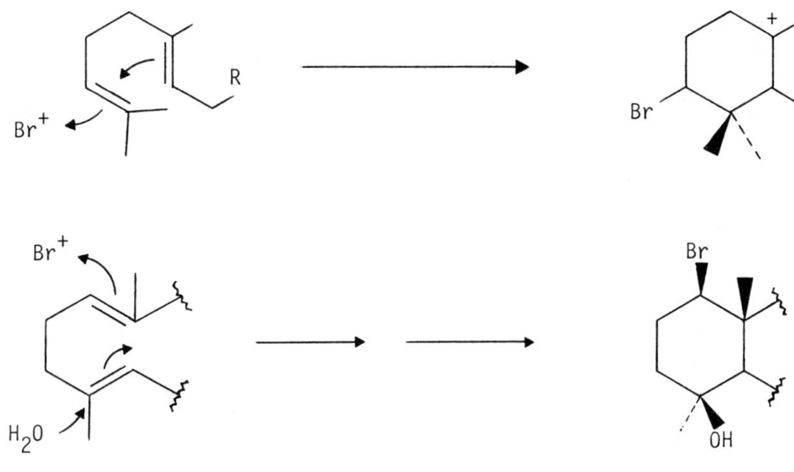

Reference: Fenical (1979).

Fig. 5.4 – Cyclization of brominated linear isoprenoids to yield a variety of terpenoids.

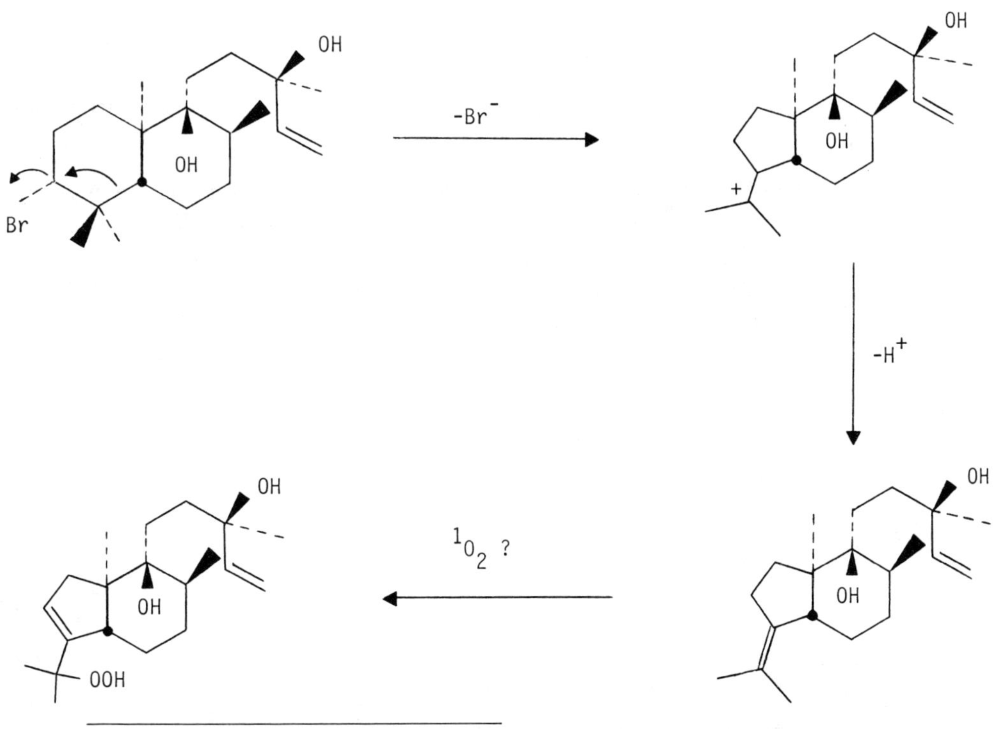

Reference: Fenical (1979).

Fig. 5.5 – Ring contraction of the brominated diterpenoid concinndiol to an unbrominated hydroperoxide metabolite.

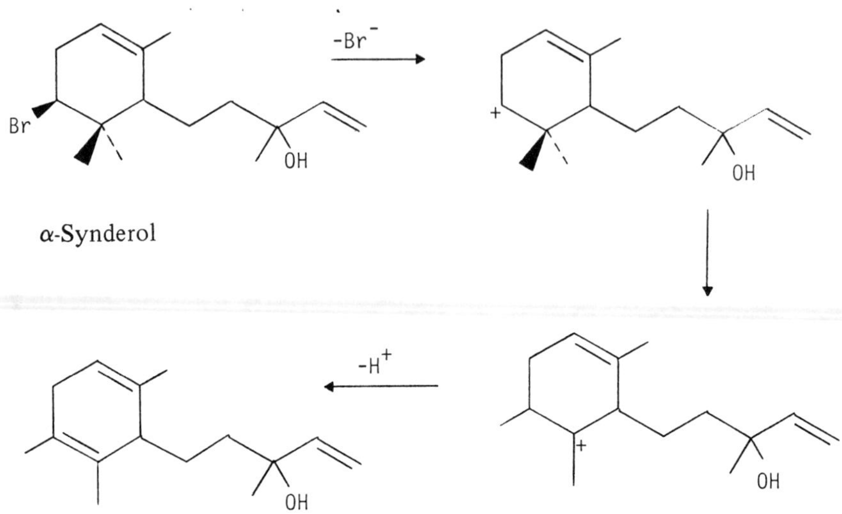

α-Synderol

Rearranged terpenoid

Reference: Fenical (1982).

Fig. 5.6 – Methyl migration in α-snyderol to yield unbrominated rearranged terpenoids.

Fenical (1982) has further suggested that bromonium ion-induced cyclizations of geraniol-derived precursors in marine organisms replaces hydrogen ion-induced cyclizations that occur in the terrestrial environment. However, even in the terrestrial realm, a model system involving an iodinium electrophile has been proposed (Johnson & Croteau 1984). This is for the conversion of α-terpinene to ascaridole (Fig. 5.7) by *Chenopodium ambrosioides* (American wormwood).

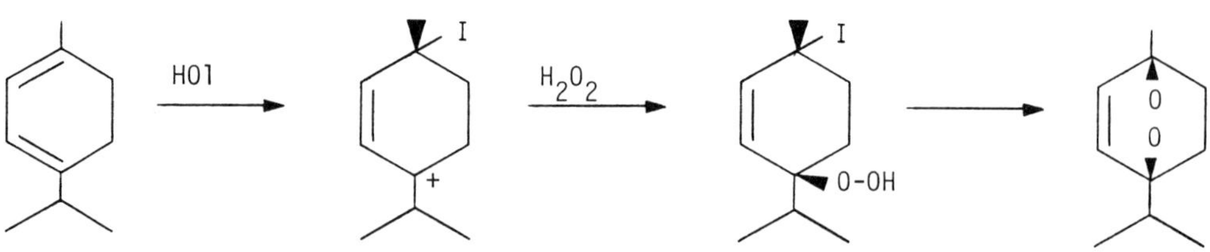

α-Terpinene

Ascaridole

Reference: Johnson & Croteau (1984).

Fig. 5.7 – Model reactions for the conversion of α-terpinene to ascaridole.

The reaction is catalyzed by an iodoperoxidase. Apparently, iodide can be accumulated from the soil by many plants, including chenopods. The authors parenthetically indicate, based on the work of others, that plants release I_2 into the atmosphere. The *in vitro* conversion of iodide to iodine by a variety of haloperoxidases, including the iodoperoxidase horseradish peroxidase, has been demonstrated (Neidleman & Geigert 1981).

HALOGENATION MAY FAVORABLY ENHANCE BIOLOGICAL ACTIVITY

The second reason why marine organisms might have a vested interest in the biosynthesis of halogenated compounds is that halogenation of organic molecules often confers high levels of biological activity even in precursors with little or no intrinsic activity. The number of pesticides that are halogenated reflect this conclusion. Antibiotics synthesized by terrestrial species, including chloramphenicol, griseofulvin, and aureomycin, contain chlorine. Kwasniewska & Kaiser (1983) have shown that chlorinated phenols are up to 500-fold more toxic to five strains of yeast than phenol itself. Toxicity increased as the number of chlorine atoms increased: 2,4,5 trichlorophenol > 2,4-dichlorophenol > p-chlorophenol.

A process for the production of antimicrobial agents has been patented (Neidleman 1974) in which whole broths, filtrates or cell extracts of microorganisms were treated with a source of positive halogen, halonium ion, (N-haloamides, N-haloimides, or a halogenating enzyme). In many cases, it was found that such biological material, having little or no intrinsic antimicrobial activity, was modified to possess substantial activity after halogenation against such pathogens as *Staphylococcus aureus*, *Escherichia coli*, and *Candida albicans*. It would appear that nature, especially in the marine environment, has reduced to practice halogenation as a mode of generating biologically active compounds.

It should be emphasized, however, that the ability of halogens to initiate or increase biological activity in organic molecules involves a number of contributing factors, including: 1) positional effects – the location of the halogen atom, 2) the number of halogen atoms, and 3) the chemical nature of the halogen atom. These points can be illustrated using, as a model, the pyrrolomycins, already noted in Chapter 2, Fig. 2.1, and Chapter 3, Fig. 3.9. The microorganisms tested for pyrrolomycin inhibition are shown in Table 5.5, and include gram positive and gram negative bacteria, as well as yeast and fungi (Ezaki *et al.* 1981). In Table 5.6 are shown inhibitory effects against this spectrum of test organisms by a selection of chloro-3-nitropyrroles, including the 4,5-dichloroderivative, pyrrolomycin A. In broad terms, the overall activity of these compounds as related to chlorine position is in the order: 4,5 > 2 > 2,5 > 5 = 2,4,5 > 2,4. The following conclusions can be reached:

1. The most active compound is the 4,5-derivative, the naturally occurring pyrrolomycin A.

2. The placement of chlorine at position 5 imparts much greater biological activity than placement at position 2, when associated with an additional chlorine at position 4.
3. The number of chlorine atoms, in the series of compounds studied, did not appear directly involved in determining the magnitude of microbial inhibition.
4. Increased activity versus gram negative microorganisms (4,5 and 2) and yeast/fungi (4,5 and 2,5) can be obtained by positional effects and number of chlorine atoms.

Table 5.5 – Test organisms for pyrrolomycin testing.

Class	Test organisms
Gram positive	*Staphylococcus aureus* 209P JC-I
	Staphylococcus epidermidis ATCC 14990
	Streptococcus faecalis ATCC 8043
	Bacillus anthracis No. 119
Gram negative	*Escherichia coli* NIHJ JC-2
	Citrobacter freundii GN-346
	Salmonella typhi 0-901-W
	Shigella sonnei EW-33 Type I
	Klebsiella pneumoniae PCI-602
	Proteus vulgaris OX-19
	Proteus morganii Kono
	Proteus mirabilis J-0013
	Serratia marcesens MB-3848
	Pseudomonas aeruginosa MB-3829
Yeast/fungi	*Candida albicans* C-A-24
	Cryptococcus neoformans Cr-I
	Trichophyton mentagrophytes 530324
	Trichophyton interdigitale
	Aspergillus fumigatus Saito

In Table 5.7, the structures of the various pyrrolomycins are repeated along with the same kind of data for microbial inhibition discussed from Table 5.6. The table is divided into three parts on the basis of structural considerations. In Table 5.7, I, is the comparison of pyrrolomycin A and B. The following observations can be made:

1. The simple compound A has greater activity against yeast and fungi.
2. The addition of the methylene dichlorophenol moiety yields increased anti-Gram positive activity while reducing yeast/fungi effects.
3. Activity against Gram negative organisms is similar for both compounds.

Table 5.6 – Structure and inhibitory activity of various chloro-3-nitropyrroles.

Microorganisms	Chlorine position:	Chloronitropyrroles					
		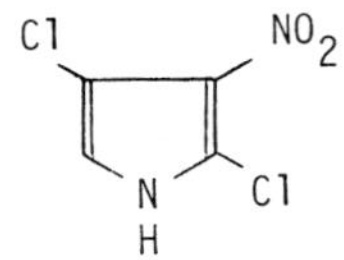	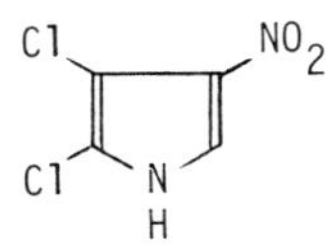	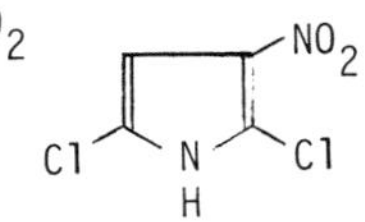	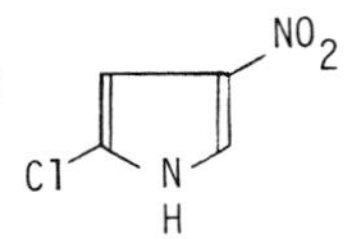	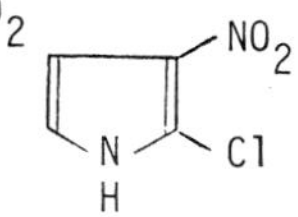	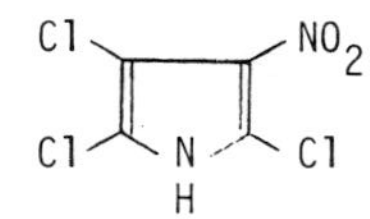
		2,4	4,5*	2,5	5	2	2,4,5
Gram + bacteria		0	++	++	+	++	+
Gram − bacteria		0	++	+	+	++	+
Yeast/fungi		+	+/++	+/++	0/+	0/+	0/+

0 = minimum inhibitory concentration >100 μg/ml.
\+ = minimum inhibitory concentration between 25–100 μg/ml.
++ = minimum inhibitory concentration between 1–25 μg/ml.
+++ = minimum inhibitory concentration <1 μg/ml.

* pyrrolomycin A.
Reference: Koyama *et al.* (1981).

Table 5.7 – Structures and inhibitory activity of various pyrrolomycins.

			Microorganisms		
		Pyrrolomycins	Gram + bacteria	Gram – bacteria	Yeast/ fungi
I.	A	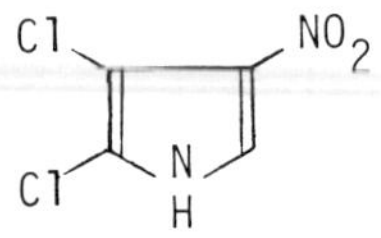	++	++	+/++
	B	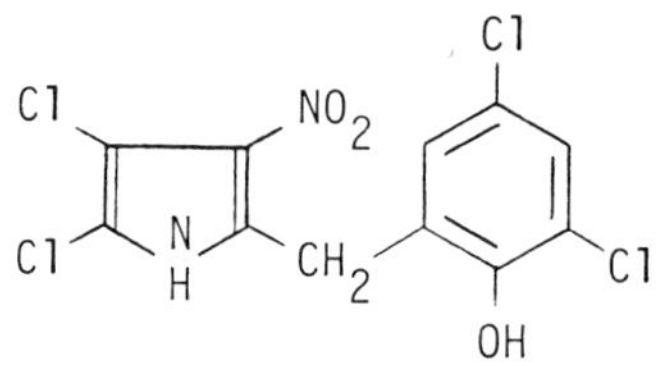	+++	++	+
II.	C	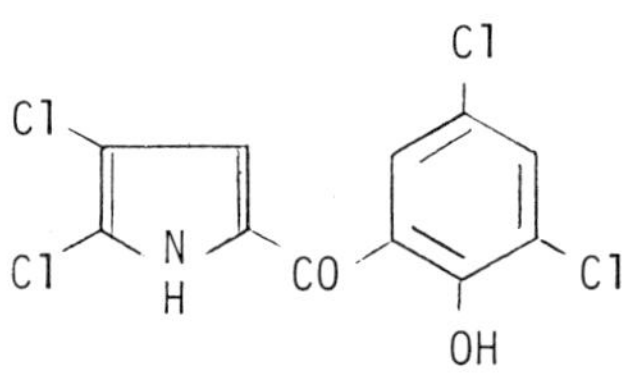	+++	0	0
	D	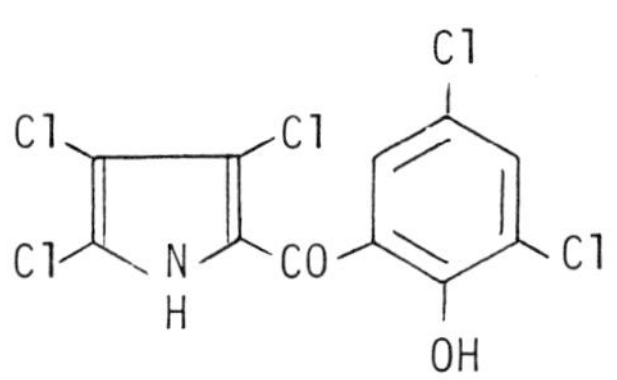	+++	++	+/++
	E	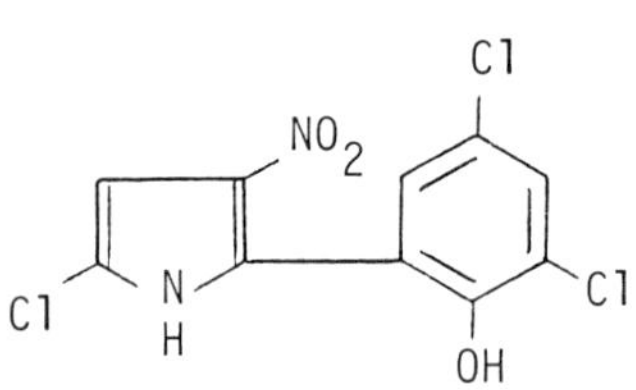	+++	0	0/+
III.	F_1	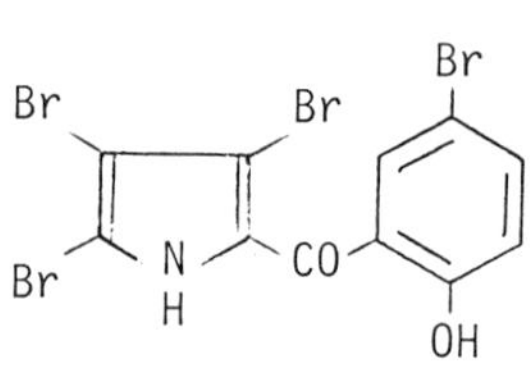	+++	++	0/+++

Table 5.7 – *Continued.*

Pyrrolomycins		Microorganisms: Gram + bacteria	Gram − bacteria	Yeast/fungi
F_{2a}	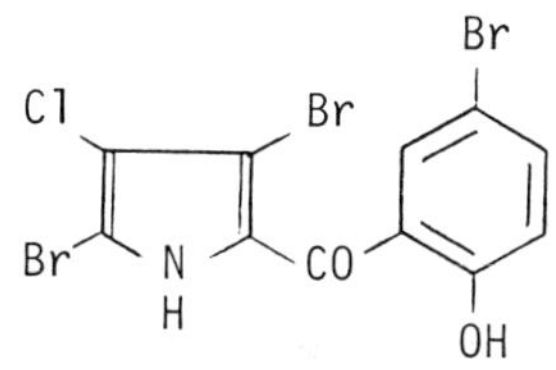	+++	few +/ mainly ++	0/+++
F_{2b}	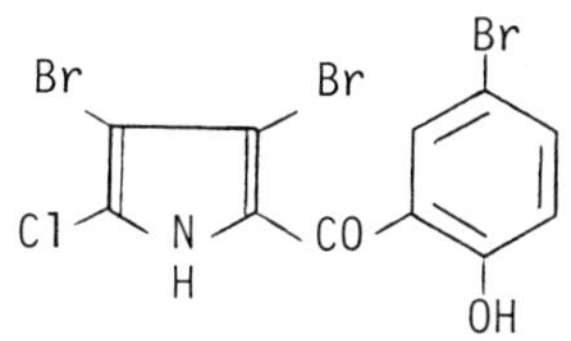	+++	mainly +/ some ++	0/++
F_3	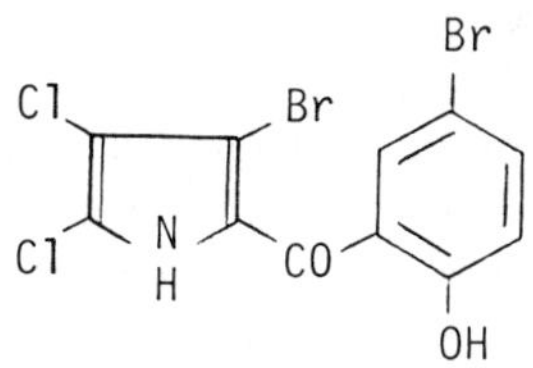	++	++	0/++/+++

0 = minimum inhibitory concentration >100 μ/ml.
\+ = minimum inhibitory concentration between 25–100 μ/ml.
++ = minimum inhibitory concentration between 1–25 μ/ml.
+++ = minimum inhibitory concentration <1 μ/ml.

References: Ezaki *et al.* 1981, Ezaki *et al.* 1983a,b.

In Table 5.7, II, the clearest result is that an additional chlorine atom in D as compared to C affords an impressive improvement in anti-Gram negative and anti-yeast/fungi inhibition. This could be either a positional or a number of chlorine atoms effect. Sufficient data to reach a conclusion are not given. It is interesting to note that, regardless of any other inhibitory activity, the anti-Gram positive effects of compounds C, D, E are equivalent. Table 5.7, III, indicates a complex set of relationships related to the chemical nature and numbers of two different halogens, chlorine and bromine. These observations can be made:

1. The general order of microbial inhibition for these compounds is: $F_1 \geqslant F_{2a} \geqslant F_3 > F_{2b}$ from Gram negative bacteria and yeast/fungi.
2. Activity against Gram positive organisms is similar for all compounds.
3. The presence of three bromine atoms yields the most active compound, whereas in compounds with both bromine and chlorine, both the number and position of a particular halogen can influence activity versus specific microorganisms.

These data suggest that by varying the number and nature of halogens in a compound as synthesized in nature and the laboratory, both the level and specificity of microbiological inhibition can be modulated. The assumption that such modulation would also be achieved in biological activities other than growth inhibition seems reasonable.

Hager (1982) reported that, in studying extracts of 1200 samples of marine microorganisms, there was a strong correlation between the antimicrobial activity of lipid cell extracts and their halometabolite content. Fenical (1974) indicated that a broad range of haloketones isolated from the red seaweed *Asparagopsis taxiformis* showed strong antimicrobial activity against a variety of microorganisms. The major compounds were 1,1,3,3-tetrabromoacetone and tribromo-3-buten-2-one. Fenical (1975) comments that in various species of red algae, halometabolites such as brominated phenols, including terpene phenols, are involved in chemical defense against parasitic microflora as well as being involved as feed deterrents against invertebrate predators. Sheikh & Djerrasi (1975) have suggested that bromophenols have a similar protective role in the marine tube worm, *Phoronopsis viridis.* Further, McConnell & Fenical (1980) have shown that the family *Bonnemaisoniaceae* produces an array of C_7-C_9 ketones, alcohols, and acids that appear to be stored in specialized vesicular cells, referred to earlier as sites of bromide concentration, for use as antimicrobial and antifeedant agents. For example, 1,3-dibromo-2-heptanone shows minimal inhibitory concentrations of 0.05 μg/ml against both *Staphylococcus aureus* and *Candida albicans.*

In the screening work reported by Hager (1982), the highest level of "lipid halometabolite concentration" was obtained from *Asparagopsis taxiformis,* 2.4 mg/g wet weight tissue. This plant produced 3–5% of its dry weight as halogenated methane derivatives, with bromoform ($CHBr_3$) being the major component. Halomethanes are considered potent biocides (Fenical 1982) and represent another class of probable defense compounds.

Other halometabolites have been demonstrated to have chemical defense functions in marine organisms. For example, among the green algae, *Avrainvillea longicaulis* has been reported by Sun and co-workers (1983) to produce the brominated diphenylamine derivative, avrainvilleol (see Chapter 2, Fig. 2.4), which has antifeedant activity. Walker & Faulkner (1981) suggest that chlorinated acetylenes such as

$$\mathrm{ClCH{=}CHCH{=}CHC{\equiv}CC{\equiv}CCH{=}CH(CH_2)_4COCH_3}$$

(1Z,3E,9Z)-1-chlorohexadeca-5,7-diyne-1,3,9-trien-15-one

are involved in the chemical defense mechanism of the Nudibranch mollusc, *Diaulula sandiegensis.*

A clever protective method has been developed by the sea hares (*Aplysia* spp.). They derive halogenated feeding deterrents from their algal diet of *Laurencia*, *Plocamium*, and *Chondria*. The algal products are concentrated in the sea hare's digestive gland and diffuse through the animal (Fenical 1975, Ireland *et al.* 1976, Kinnel *et al.* 1977, 1979, Thomson 1978, Dieter *et al.* 1979, Cardinella *et al.* 1982, Wood 1983, Naylor *et al.* 1983). Some of these compounds are illustrated in Fig. 5.8. It should be emphasized that not all feed-deterrent compounds are halogenated (Minale & Riccio 1976, Sun & Fenical 1979, Paul *et al.* 1982).

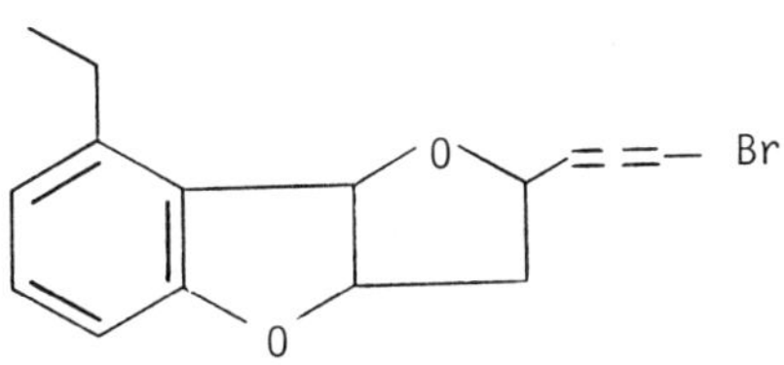

panacene

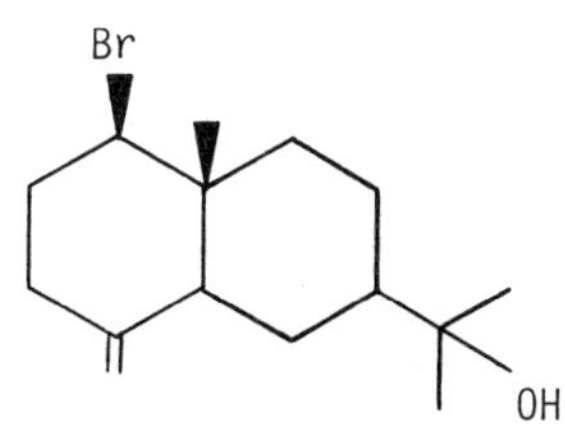

brasudol

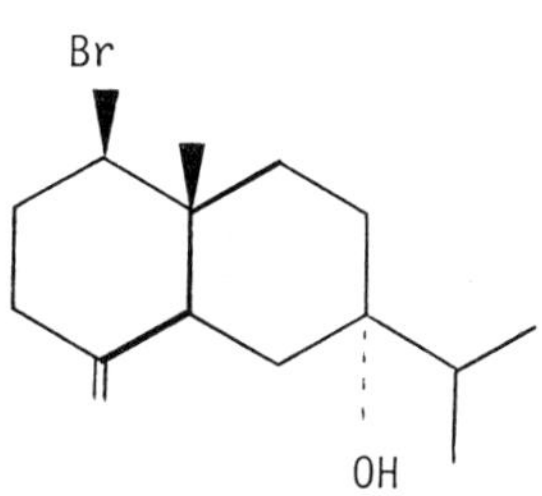

isobrasudol

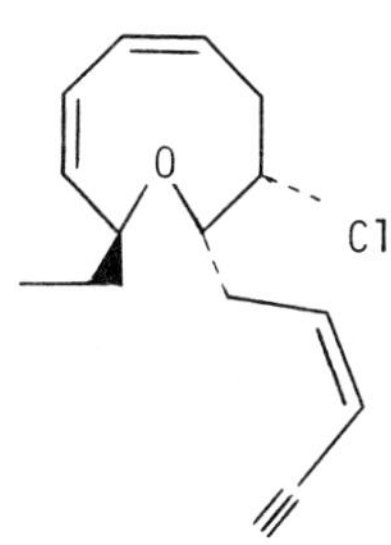

brasilenyne

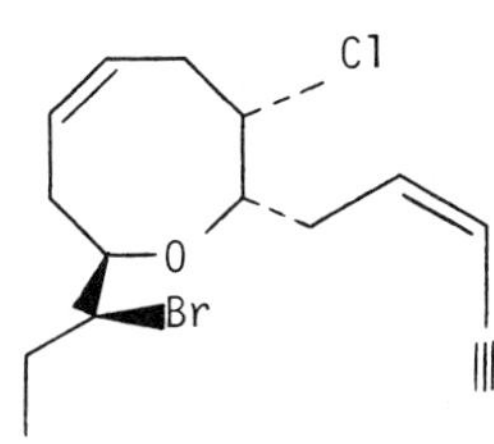

cis-dihydrorhodophytin

Fig. 5.8 – Feeding deterrent compounds isolated from *Aplysia* spp.

Iodinated tyrosines including triiodothyronine and thyroxine have been reported to occur in red and brown algae (Ragan 1981). Do these compounds exert a metabolic effect in these algae or their predators similar to that of thyroid hormone in mammals?

A striking example of the possible interplay between biological activity in marine halometabolites and medical research is illustrated in the study of the pair of algal metabolites in Fig. 5.9. Using a tumor promoter sensitive and transformable mouse cell clone, Shimomura and co-workers (1983) have shown that whereas aplysiatoxin increased malignant transformation, stimulated DNA synthesis, and inhibited the binding of phorbol-12,13-dibutyrate and epidermal growth factor to cell receptors, the debromo-compound carried out only the binding inhibition reactions. Owing to the dramatic difference in the biological activities of the two compounds, they are valuable in studies related to tumor promotion and the effect of the hydrophilic region of the inhibitor (in which the bromine atom is located) on the process. Conversely, are these differences in the biological activities of these two substances indicative of qualitatively diverse roles in the physiology of *Lyngbya gracilis*? The same pair of compounds is the toxin of the sea hare, *Stylocheilus longicauda*, another example of the acquisition of defense chemicals obtained by the sea hare feeding upon seaweeds (Faulkner 1977).

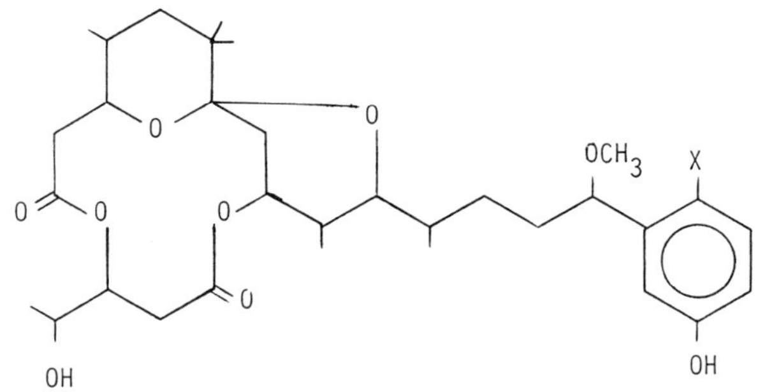

X = Br in aplysiatoxin

X = H in debromoaplysiatoxin

Fig. 5.9 – Structures of aplysiatoxin and debromoaplysiatoxin from the alga, *Lyngbya gracilis.*

A further example of a halogenated, marine metabolite with implications for mammalian physiology is the compound autonomium, isolated from the sponge, *Verongia fistularis.* The structure is shown in Fig. 5.10 along with that of acetyl choline and adrenaline. It can be seen that autonomium shares some of the chemical attributes of both acetyl choline and adrenaline. In addition, autonomium shares the pharmacological properties of these two compounds:

α- and β-adrenergic and cholinergic activities (Kaul 1982). Is autonomium a primitive precursor of adrenaline and acetyl choline, or is it an indicator that such dual activity compounds may exist in mammals? What is the role of autonomium in the producing organism?

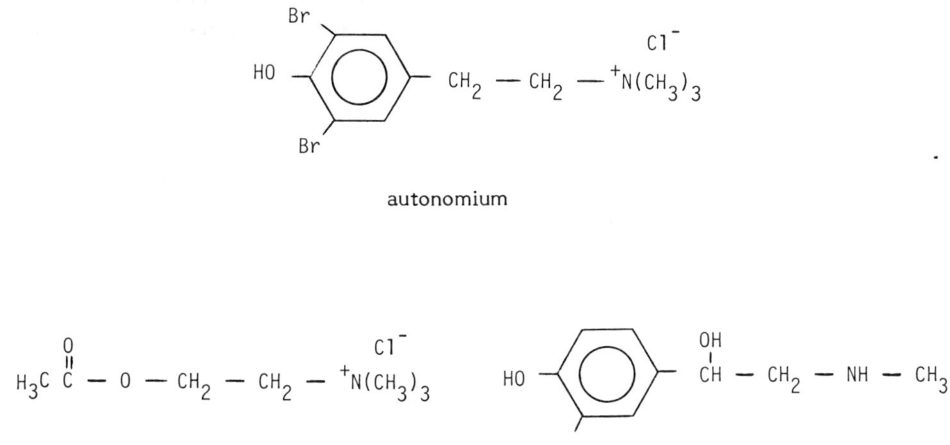

autonomium

acetylchlorine

adrenaline

Fig. 5.10 – Structural relationship of the marine metabolite autonomium to adrenaline and acetyl choline.

A final illustration of halogenated marine products with pharmacological activity in mammals is that of the cyclic ethers, dactylyne and laureatin, isolated from the sea hare, *Aplysia dactylomela*, shown in Fig. 5.11. These compounds potentiate the phenobarbital-induced hypnosis in rats and mice by apparently inhibiting phenobarbital degradation by cytochrome P_{450} (Kaul 1982). These, and related compounds showing similar activity, have in common the conjugated eneyne exocyclic side chain. What physiological function do these compounds have for the sea hare?

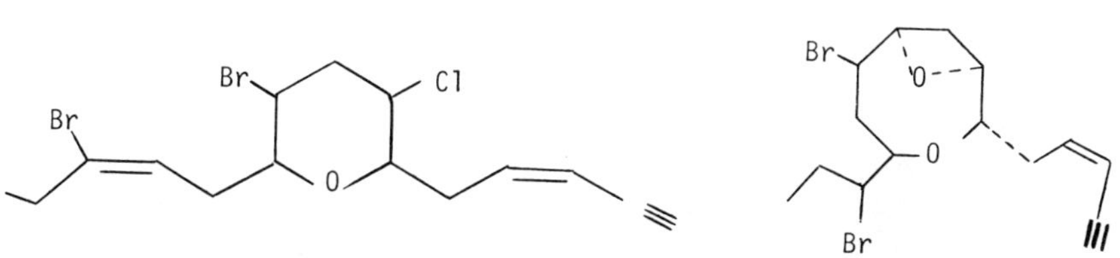

dactylyne

laureatin

Fig. 5.11 – Marine metabolites potentiating phenobarbital activity.

Chapter 6, on the medical significance of the halogenating enzymes, will detail information on the role that halogenating enzymes play in the defense mechanisms of the human. One of the questions that may be raised in relation to the role of these enzymes in marine organisms is whether or not they play a more direct role in destroying predators, as do myeloperoxidase, lactoperoxidase, and eosinophil peroxidase in the human. Do these marine enzymes generate active halogenating species such as chloramines, or pseudohalogen species such as HOSCN that inactivate critical chemical or enzymatic components in predators, thereby destroying them?

6

Halogenating enzymes and the mammalian defense mechanism

THE ENZYMES AND THEIR LOCATION

It is clear that mammalian halogenating enzymes are directly involved in the body's defense against invading microorganisms, parasites, and, perhaps, tumor cells and other cell types. The most studied enzymes are myeloperoxidase (MPO) and lactoperoxidase (LPO), although some work has also been done with the peroxidase of eosinophils (EPO). MPO is located in the polymorphonuclear leukocyte (PMN), and LPO is detected in human saliva, milk, and tears (Schultz 1980, Thomas *et al.* 1981, Clark 1983, Smith & Yang 1984). While MPO appears to be ubiquitous in location owing to the travels of the PMN, LPO has the oral cavity as its major site of activity.

THE MECHANISM OF THE TOXIC REACTIONS

The broad spectrum of substrates that can be reacted upon by the halogenating enzymes was indicated in Chapter 3. It is clear that if biologically relevant substrates, related structurally to these model substrates and vital to the survival of living cells, are inactivated by enzymatic modification, or if the product of the reaction between the enzyme and the biologically relevant substrate has toxic properties, then the enzymes will have potential for the destruction of pathogenic cells.

In the presence of a halogenating enzyme, H_2O_2, and an appropriate halide ion or thiocyanate ion, halogenation, thiocyanation, or oxidation can occur.

Whereas MPO and EPO utilize chloride ion (iodide ion to a lesser degree), LPO utilizes thiocyanate ion, primarily. MPO, which can accompany LPO in the oral cavity, can also use thiocyanate ion.

The MPO and EPO systems produce a variety of reactive substances including halogens, hypohalous acids, and chloramines (Albrich *et al.* 1981, Albrich & Hurst 1982, Clark 1983, Foote *et al.* 1983, Thomas *et al.* 1983a, Weiss *et al.* 1983). LPO and MPO produce hypothiocyanous acid (HOSCN) (Thomas *et al.* 1981, Tenovuo *et al.* 1982). It is likely that these agents are most responsible for the toxic effects on target cells and molecules. It should be noted that, in addition to halide-dependent toxicity, MPO, in particular, has been associated with O_2-dependent toxicity, mediated through superoxide anion, hydroxyl radical, or singlet oxygen (Clark 1983, Kanofsky *et al.* 1984).

A more detailed picture of the mechanisms involved in the cytotoxic effects of halogenating enzymes will be obtained below when biochemical or molecular targets of the enzymes are illustrated, after a brief consideration of the cellular targets.

THE CELLULAR TARGETS OF THE HALOGENATING ENZYMES

Cellular targets of MPO and EPO are killed either intracellularly within a phagocytic vacuole as in the case of bacteria or extracellularly as in the case of mammalian cells, hyphal forms of fungi, and multicellular metazoan parasites. With LPO, since it is an enzyme secreted from exocrene glands into saliva, milk, and tears, killing is extracellular.

Table 6.1 indicates some *in vitro* antimicrobial activities that have been demonstrated with MPO. It can be seen that representatives of Gram positive and Gram negative bacteria, as well as fungi and yeast, are enumerated. The *in vivo* activity of MPO has been studied by Wright & Nelson (1985).

Table 6.1 – *In vitro* antimicrobial activity of myeloperoxidase.

Inhibited microorganism	Reference
Aspergillus fumigatus	Diamond & Clark (1982)
Bacillus cereus	Tenovuo *et al.* (1985)
Candida albicans	Diamond *et al.* (1980)
Escherichia coli	Klebanoff (1967) (also with EPO, Jong, *et al.* 1980)
Klebsiella pneumoniae	Tsan (1982b)
Rhizopus oryzae	Diamond & Clark (1982)
Staphylococcus aureus	Tsan (1982b)
Streptococcus mutans	Kersten *et al.* (1981)

Table 6.2 shows similar information for LPO. Most of the bacteria listed occur in the human oral cavity or in milk, the two prime sites of LPO activity.

It is believed that LPO in the presence of thiocyanate ion mediates its effects by reactions with sulfhydryl (–SH) groups in proteins (Thomas *et al.* 1983b). In strains of *Streptococci*, Carlsson and co-workers (1983) have indicated that the glycolytic pathway and glyceraldehyde-3-phosphate dehydrogenase in particular appear to be the targets of LPO.

Table 6.2 – *In vitro* antimicrobial activity of lactoperoxidase.

Inhibited microorganism	Reference
Bacillus cereus	Zajak *et al.* (1981)
	Tenovuo *et al.* (1985)
Escherichia coli	Thomas & Aune (1978a)
	Marshall & Reiter (1980)
Pseudomonas fluorescens	Reiter & Harnulv (1982)
Salmonella typhimurium	Purdy *et al.* (1983)
Streptococcus agalactiae	Mickelson (1979)
Streptococcus lactis	Marshall & Reiter (1980)
Streptococcus mitis	Carlsson *et al.* (1983)
Streptococcus mutans	Carlsson *et al.* (1983)
	Thomas *et al.* (1983b)
Streptococcus salivarius	Carlsson *et al.* (1983)
Streptococcus sanguis	Carlsson *et al.* (1983)

In Table 6.3 are illustrated some of the mammalian cell types against which MPO shows *in vitro* cytotoxic activity. Also noted are two parasites against which EPO is effective.

Table 6.3 – *In vitro* cytotoxicity activity of myeloperoxidase and eosinophil peroxidase.

Target cells and parasites	Enzyme	Reference
erythrocytes	MPO	Klebanoff & Clark (1975)
fibroblasts, human	MPO	Clark (1983)
leukocytes	MPO	Clark & Klebanoff (1977)
murine lymphoma (L1210)	MPO	Edelson & Cohn (1973)
murine lymphoma (LSTRA)	MPO	Clark *et al.* (1975, 1976)
murine lymphoma (LSTRA)	EPO	Jong & Klebanoff (1980)
platelets	MPO	Clark & Klebanoff (1980)
Schistosoma mansoni schistosomula	EPO	Jong *et al.* (1981)
spermatoza	MPO	Smith & Klebanoff (1970)
Trichinella spiralis larvae	EPO	Buys *et al.* (1981)

The data summarized in these tables indicate the broad cytotoxic and antimicrobial effects affected by halogenating enzymes *in vitro*. Research into the *in vivo* significance of these observations is slowly developing.

An effect more subtle than cytotoxicity is the reported suppression of natural killer (NK) activity of mononuclear leukocytes by MPO, H_2O_2, and Cl^- (El-Hag & Clark 1984). The viability of the leukocytes was >95% despite the suppression of NK activity.

BIOCHEMICAL AND MOLECULAR TARGETS OF HALOGENATING ENZYMES

In Chapter 3 it was discussed that compounds containing sulfur, amino groups, phenol rings, and double bonds were substrates for halogenating enzymes. These general conclusions apply as well to biologically significant molecules such as proteins, peptides, amino acids, and carotenoids.

Table 6.4 lists a number of nonpolymeric biomolecules that have been shown to be modified by MPO. It is assumed that most of these molecular alterations are either directly or indirectly detrimental to cell survival and represent, to some degree, the mechanism of cytotoxicity and antimicrobial activity elicited by halogenating enzymes. A direct effect, for example, is the oxidation of methionine to methionine sulfoxide. An indirect effect is the formation of chloroamines of amino acids and other amines, which, in turn, react with other molecular species.

In Table 6.5 are peptides and proteins that have been shown to be inactivated by MPO. It can be seen that these polymeric substances are susceptible to MPO in much the same way as the simple compounds of Table 6.4. It should be noted that methionine is a particularly sensitive and important target both as a free amino acid and as a component within a peptide or protein. The biological significance of these enzymatic alterations remains to be seen, in many cases. However, in the instance of the leukotrienes, Lee and co-workers (1983b) suggest that the modifications have implications for regulation of the inflammatory process.

Not every instance wherein MPO and halogenation are implicated results in biological inactivation. For example, Weiss and co-workers (1985) have reported the oxidative autoactivation of latent collagenase of neutrophils involving production of hypochlorous acid by these same cells. This represents another involvement of MPO in the inflammation process since collagenase, so activated, is involved in the destruction of collagen that occurs during the inflammatory process.

MEDICAL SIGNIFICANCE OF HALOPEROXIDASE ACTIVITY

Aside from any direct contributions MPO makes to the defense mechanism of mammals, there are other positive and negative biological consequences that follow from the broad spectrum of chemical reactivity possessed by primary and secondary reaction products of the enzymes. Two scenarios illustrate this point.

Table 6.4 – Nonpolymeric biochemical targets of myeloperoxidase.

Target	Susceptible functionality	Reaction of product	Reference
aflatoxins	–	oxidation	Odajima (1981)
amino acids: L-alanine β-alanine, glycine, taurine, etc.	amino	chloramine	Selvaraj *et al.* (1978) Thomas *et al.* (1982, 1983a,b)
		decarboxylation	Tsan (1982b) Weiss *et al.* (1982, 1983) Cramer *et al.* (1984)
β-carotene	double bonds	halogenation, oxidation	Kanner & Kinsella (1983)
methionine	$-SCH_3$, carboxyl	sulfoxide, decarboxylation	Tsan & Chen (1980) Tsan (1982a, b)
NAD(P)H	ring N	chloramine	Selvaraj *et al.* (1980)
tryptophan and tyrosine	ring system	iodinated	Schonbaum (1976), Thomas & Aune (1978b) Alexander (1974), Morrison & Schonbaum (1976), Thomas & Aune (1978b)

Table 6.5 – Polymeric biochemical targets of myeloperoxidase.

Target	Susceptible functionality	Reaction of product	Reference
acid phosphatase	amino, sulfhydryl, etc.	–	Kobayashi, *et al.* (1982)
α-amylase (barley)	tyrosine	chlorinated	Mitchell *et al.* (1981)
α-amylase (*B. subtilis*)	methionine	sulfoxide	Mitchell *et al.* (1981)
N-acetyl-β-glucosaminidase	amino, sulfhydryl, etc.	–	Kobayashi *et al.* (1982)
bovine serum albumin	sulfhydryl	disulfide	Mitchell *et al.* (1981)
complement component C3b	sulfhydryl	–	Coble *et al.* (1984)
E. coli cell envelope	–	–	Sips & Hamers (1981)
E. coli proteins	amino, carboxyl	–	Paul *et al.* (1970), Selvaraj *et al.* (1974)
α-fucosidase	amino, sulfhydryl, etc.	–	Kobayashi *et al.* (1982)
β-glucosaminidase	amino, sulfhydryl, etc.	–	Kobayashi *et al.* (1982)
β-glucuronidase	amino, sulfhydryl, etc.	–	Kobayashi *et al.* (1982)
glutathione	sulfhydryl	disulfide	Turkall & Tsan (1982)
heme and non-heme iron proteins	Fe centers	Fe loss	Rosen & Klebanoff (1982)
IgG (no halide involved)	–	aggregates	Jasin (1983)
immunoglobulin G	sulfhydryl	–	Coble *et al.* (1984)
β-lactoglobulin	sulfhydryl	disulfide	Thomas & Aune (1977, 1978b)
leukoattractant peptides (bacterial and human complement)	methionine	sulfoxide	Clark *et al.* (1980) Clark (1982, 1983) Lane & Lamkin (1983)
leukotrienes C_4, D_4, E_4	glutathione	sulfoxide	Lee *et al.* (1983b)
lysozyme	amino, sulfhydryl, etc.	–	Kobayashi *et al.* (1982)
α-mannosidase	amino, sulfhydryl, etc.	–	Kobayashi *et al.* (1982)
α_1-proteinase inhibitor	methionine	sulfoxide	Matheson (1982), Clark (1983)
slow reacting substance of anaphylaxis (SRS-A)	thioether	oxidized	Hendersen *et al.* (1981)

Leukocyte chemotactic factors or leukoattractant peptides such as indicated in Table 6.5 have in common the presence of methionine. When a 'sufficient' population of MPO-containing PMN have been attracted by chemotactic peptides to a site of pathogen invasion or injury, these peptides induce secretion of MPO and hydrogen peroxide, which, in the presence of chloride ion, cause oxidation of the peptidyl methionine to methionine sulfoxide and inactivation of the leukoattractant peptide. This results in modulation of the binding of the chemotactic peptides to receptor sites of the PMN, thereby modulating chemotactic events in the inflammatory loci. In this feedback reaction, the peptide both attracts the PMN and causes its own degradation (Clark 1982, 1983, Lane & Lamkin 1983).

A second but negative scenario involves the α_1-proteinase inhibitor. Usually, this plasma proteinase inhibitor exerts control over connective tissue proteolysis such as that mediated by neutrophil elastase. In pulmonary emphysema and rheumatoid arthritis, a massive influx of PMN and, therefore, of MPO occurs and results in the inactivation of the α_1-proteinase inhibitor, allowing tissue injury to occur. A methionine residue critical to binding the inhibitor to the proteinase is oxidized to the sulfoxide, thus inactivating the inhibitor (Travis, *et al.* 1984).

D-penicillamine is used in the treatment of rheumatoid arthritis. Cuperus and co-workers (1983) have shown that the compound, depending upon its concentration, can inhibit HOCl formation by MPO and also the oxidizing activity of HOCl. In the latter instance, D-penicillamine is itself a substrate for oxidation by HOCl yielding D-penicillamine disulfide, penicillamine sulfinic acid, and a deaminated product. This multistep degradation allows one molecule of penicillamine to scavange at least five molecules of HOCl. In inhibiting MPO, D-penicillamine stabilizes compound II, an inactive form of the enzyme. This effect of penicillamine inhibiting halogenation (oxidation) is reminiscent of the effect of benzothiazole on chlortetracycline formation discussed in Chapter 3.

By way of contrast, Bolscher and co-workers (1984b) have reported that ascorbic acid (vitamin C) has a stimulating effect on the chlorinating activity of MPO by increasing the rate of conversion of inactive compound II to native MPO. They suggest that this activity may be of physiological importance. Since elevated MPO activity may be beneficial by affording increased antimicrobial activity or detrimental by inhibiting the action of α_1-proteinase and other plasma inhibitors, the presumed physiological effect of vitamin C on MPO activity may be good or bad, depending upon the particular medical situation.

It is clear that such broadly reactive enzymes as the halogenating enzymes are both good news and bad, and must be controlled as strictly as possible by the producing organism. In this regard, it is interesting to note that it has been reported that the LPO-H_2O_2-thiocyanate system does not damage DNA as determined in tests using mutagen-sensitive *Salmonella typhimurium*, nor did it appear to oxidize calf thymus DNA (White *et al.* 1983). A provocative question is whether this is equally true of the MPO-H_2O_2-chloride system. That the pyrimidine nucleus can be biologically halogenated has been confirmed by the isolation of 4-amino-5-bromopyrrolo(2,3-d)pyrimidine from a marine sponge and 5-iodo-5′-deoxytubericidin from a red alga (Kazlauskas *et al.* 1983). Such observations as these suggest the possibility that halogenated or oxidized DNA derivatives might be produced by halogenating enzymes under certain conditions, giving rise to mutagenic effects. In this regard, Weitzman and co-workers (1985) have demonstrated that malignant transformation of C3H mouse fibroblasts was caused by the presence of human neutrophils. The carcinogenic agent was not identified, but it was noted that one or more of the following enzymatic products could be responsible: superoxide anion, hydrogen peroxide, hydroxyl radical, hypohalous acids, or haloamines.

MISCELLANEOUS REACTIONS

Since the substrate specificity of halogenating enzymes is low, it is to be expected that many unanticipated substrates must exist. These substrates or their products might affect, for example, the contribution to defense mechanism by these enzymes in either a positive or negative fashion. It has been reported, for example, that the drugs shown in Table 6.6 inhibit the MPO-H_2O_2-chloride system (Pekoe *et al.* 1982, 1983), and that a primary mechanism in their therapeutic function is the inhibition of MPO activity. From the structure of the drugs, it is readily seen that each could be a substrate for chlorination or oxidation by the MPO-H_2O_2 system, and that, in part at least, this could explain their interference with MPO-catalyzed reactions, as in the case of D-penicillamine.

In a study of the effect of dimethylsulfoxide (DMSO) on the bactericidal activity of PMN, Repine and co-workers (1981) have shown that DMSO inhibits this activity. They postulate that DMSO may accomplish this by scavenging an oxidizing intermediate such as hydroxyl radical.

Geigert and co-workers (1983f) have shown that the chloroperoxidase (CPO) of *Caldariomyces fumago* can oxidize DMSO to dimethylsulfone in the presence of hydrogen peroxide and in the absence of halide ion. Neither LPO nor horseradish peroxidase carried out this reaction. Most interestingly, in the presence of halide ion, halosulfoxides and halosulfones were formed. It was suggested that these halogenated derivatives might be toxic. Further, there is the possibility that DMSO may inhibit the bactericidal activity of PMN by being an active halide or halonium ion scavenger.

CONCLUSIONS

The ability of MPO, LPO, and EPO to carry out a broad range of halide-dependent cytotoxic and antimicrobial activities, affecting a wide variety of mammalian and microbial cells, has been discussed. In addition, it was indicated that these enzymes can inactivate a broad spectrum of molecular species important to the inflammatory response, prevention of tissue proteolysis, and immunological reactions. Much of the research described has been performed using *in vitro* model systems. Just how these results relate to the *in vivo* situation is yet to be demonstrated in most cases. The next few years will see striking discoveries in this area.

Table 6.6 – Various drugs as potential substrates for the MPO-H_2O_2-chloride system.

Drug	Structure	Activity
ketamine	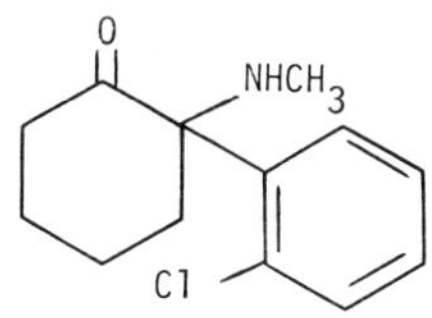	anesthetic
indomethacin	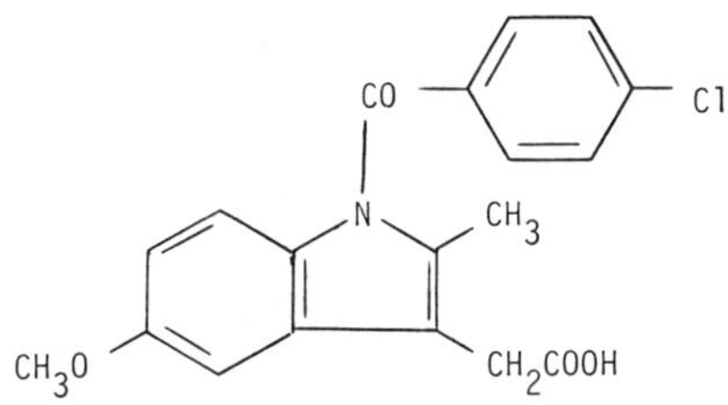	antiinflammatory
sulindac sulfide	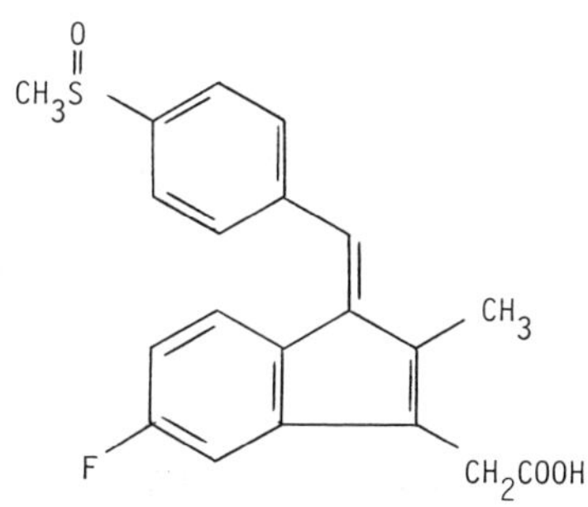	antiinflammatory
ibuprofen	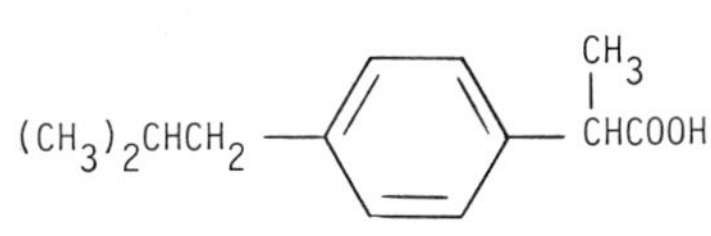	antiinflammatory

Many of the observations made in these mammalian studies seem to indicate that the halogenating enzymes in more primitive life forms such as the microorganisms, algae, and sponges in the marine environment may have similar roles, more subtle and sophisticated than those so far demonstrated. On the other hand, the synthetic activities of the halogenating enzymes, presumed to occur in primitive marine forms, lead one to ask whether there may be a similar biosynthetic function in marine and terrestrial mammals as well.

7

Commercial applications for haloperoxidases

INTRODUCTION

Enzymes such as hydrolases and isomerases have been used commercially for many years (e.g., the uses of proteases in detergents for removing blood stains and the use of glucose isomerase to convert corn syrup to high fructose corn syrup). Other enzyme classes, including the haloperoxidases, have very few commercial applications at present. Martyn and co-workers, in 1981, made the following predictive comment: '. . . the use of chloroperoxidase or other enzymatic halogenations presents almost unlimited possibilities for the production of useful chemicals in the future.' Now, four years into that future, is the same glowing evaluation of the commercial potential for haloperoxidases still valid?

PROCESS CONSIDERATIONS

Intrinsic to enzymatic processes in general are a number of technological and operational problems requiring satisfactory solutions. For haloperoxidases, two of the major problems are: 1) obtaining adequate supplies of the desired biocatalyst at the required degree of purity, and 2) optimizing operational conditions to yield economically useful enzyme activity and lifetime.

As noted in Chapter 3, many sources exist for haloperoxidases, and various isolation schemes have been developed to obtain milligram amounts of these enzymes. However, for commercial applications, grams, pounds, or even larger quantities will be needed, and producing at this level is a different matter.

Fortunately, the tools for doing this are, or are becoming, available. Such technologies include the application of molecular biology, the use of the well-tested screens for superior enzyme producers, and the improved capabilities of modern-day fermentations. As one example, Toyobo Company of Japan has developed a cell culture method for producing horseradish peroxidase (HRPO), to replace the need to grow the horseradish plant in the field. Their method is reported to give a 5–7 times higher yield of HRPO on a per gram of biomass basis than that obtained from the field-grown plant. Commercial production of HRPO through cell culture is predicted to start in 1986, with a target annual production of 10–50 tons (*Practical Biotechnology*, August 1984, p. 1). In the future, the application of recombinant DNA methodology should allow the transfer of the haloperoxidase gene from any source, for example a mammalian myeloperoxidase, to a more suitable host such as the bacterium *E. coli*, where protein overproduction readily gives yields over 20 mg of enzyme/liter in 1000-liter fermentations. For downstream processing, studies are under way to develop techniques to isolate such large enzyme quantities from crude extracts. Chavez & Flarkey (1984) have found that column packings adaptable for preparative HPLC systems can be used for haloperoxidases. Hasegawa & Kobayashi (1983) used precipitation and ion exchange chromatography, steps commonly used in industrial processes, to purify myeloperoxidase from blood.

In Chapter 4 we saw that haloperoxidases can be tuned to their maximal catalytic velocity by carefully balancing the pH and the levels of H_2O_2 and halide ion in the reaction mixture. Further enhancement of this velocity is possible by additives or solvent changes. For example, the addition of vitamin C to a myeloperoxidase-catalyzed reaction results in a 3-fold increase in the chlorination rate (Bolscher *et al.* 1984b), and the entrapment of horseradish peroxidase in hydrated, reverse micelles of surfactant/octane/water results in a 10- to 100-fold increase in activity (Martinek *et al.* 1982, Klyachko *et al.* 1984). Haloperoxidases retain significant activity in nonaqueous solvent systems, such as 10% methanol or dimethyl sulfoxide (Cooney & Hueter 1974). Even though the haloperoxidases are not organic solvent soluble, they can be made so by chemical modification, as was done with horseradish peroxidase by coupling to polyethylene glycol (Takahashi *et al.* 1984). But maximal catalytic velocity is only part of the concern in commercial applications. It must be associated with efficient utilization of reactants. For haloperoxidases, a major obstacle is the multiple mechanism pathway available to these enzymes. As described in Fig. 4.5, the enzymes can either halogenate or oxidize the organic substrate. In the absence of added halide ion, peroxidation or oxidation will occur. If sufficient halide ion is present at the start of the reaction, the enzyme will switch from halogenation to peroxidation as the halide ion is depleted during the course of the reaction. An excess of halide ion can result in different products (e.g., dihalogenation vs. monohalogenation). A deficiency of organic substrate and the enzyme will be attacked by its own intermediate product, the hypohalous acid. An excess of H_2O_2 can result in either enzyme inactivation or the rapid decomposition of hypohalous acid. These multiple and competing pathways are summarized in Fig. 7.1. Despite this complex of factors, control is possible, but with a compromise between maximal catalytic activity and efficent utilization of reactants. Fortunately, there are ways to minimize the cost of the reactants such as the use of chloride ion from seawater and the use of dilute, impure solutions of H_2O_2.

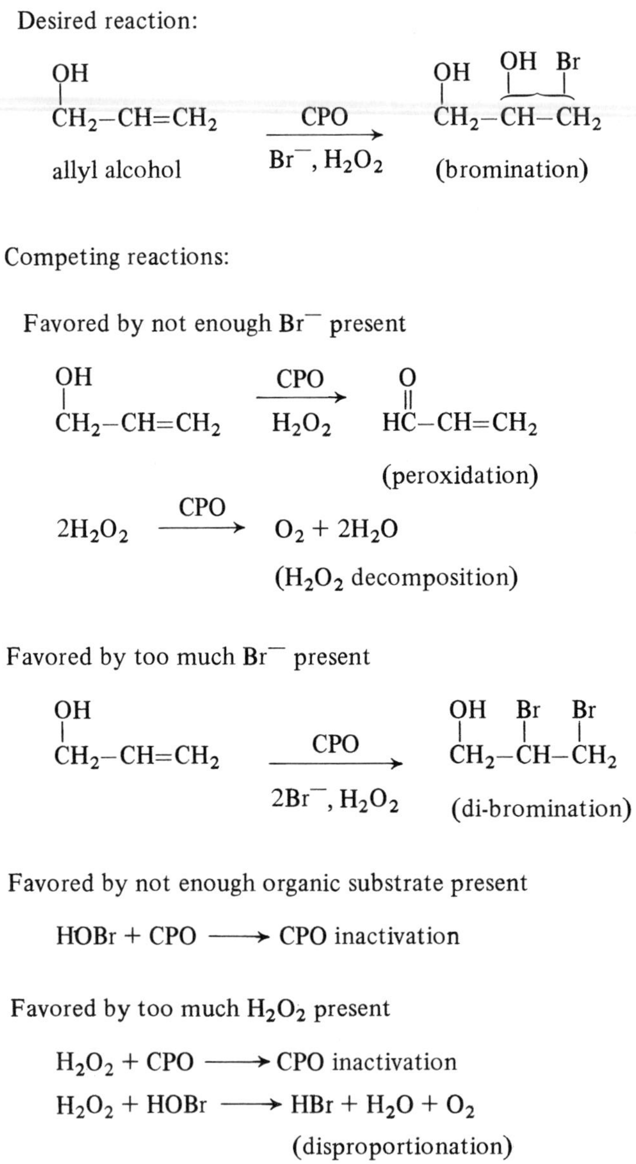

Fig. 7.1 – Multiple pathways of interaction in the haloperoxidase-catalyzed reaction – example of allyl alcohol, Br^-, H_2O_2 and chloroperoxidase (CPO).

Fast, efficient utilization of reactants and an extended operational lifetime are among the ultimate requirements for the commerical use of an enzyme. Haloperoxidases can commit suicide (i.e., attack of the enzyme-produced hypohalous acid on the enzyme itself, leading to inactivation), and this presents a major obstacle to extended operation with these enzymes. Three approaches have been used to try to control this. The first approach is the easiest, and that is to keep the organic substrate in excess in the reaction mixture so that the hypohalous acid will attack it and not the enzyme. A novel method has been used successfully for gaseous alkenes like propylene. The gaseous substrate is bubbled through the reaction mixture to maintain the necessary level of organic substrate. The halogenated product, a water-soluble liquid, remains in the reaction mixture, while the unused gaseous substrate is recycled through the reaction mixtures (Neidleman *et al.* 1981). A second approach is to control the rate of hypohalous acid generation by controlling the level of H_2O_2 present in the reaction mixture. One useful method is the slow and rate-limiting *in situ* generation of H_2O_2. This results in more than a 6-fold increase in enzyme half-life (Neidleman *et al.* 1980, 1981). A modification of this approach is to use a bioelectrochemical cell, placing glucose oxidase in the anode compartment and haloperoxidase in the cathode compartment. Control of H_2O_2 generation is regulated by the level of glucose added to the anode compartment. The current passing through the circuit to the cathode compartment is proportional to the level of glucose. At the cathode compartment, O_2 is converted to H_2O_2, which then is available for the haloperoxidase. Extended operational lifetimes for both the haloperoxidase and the oxidase were obtained (Laane *et al.* 1984, 1985). A third approach is to discover a haloperoxidase less sensitive to H_2O_2 or hypohalous acid. Nature can teach us some lessons. One of the chloroperoxidases isolated from a dematiaceous hypomycetes is orders of magnitude more resistant to hypohalous acid than any other reported haloperoxidase (Table 7.1). This

Table 7.1 – Chloroperoxidase resistance to HOCl inactivation.

Chloroperoxidase	(HOCl)*	Remaining enzyme activity
CPO from *Caldariomyces fumago*	0	100%
	50 μM	25%
	500 μM	<2%
	880 μM	0%
	4280 μM	0%
non-heme enzyme from dematiaceous hyphomycete	0	100%
	50 μM	>98%
	500 μM	>90%
	880 μM	56%
	4280 μM	32%

* 2 minute exposure time.
Reference: Geigert *et al.* (1985).

enzyme does not contain heme, which may account for its enhanced stability. One wonders if the non-heme algal iodoperoxidase and the non-heme lactoperoxidase mentioned in Chapter 3 may also have enhanced resistance to hypohalous acid attack. In addition, since it is known that hypohalous acid attacks the heme moiety and certain amino acids in the glycoprotein portion of these enzymes, it may be possible by chemical modification or site-specific mutation to impart greater resistance.

Immobilization of haloperoxidases is another way to increase the lifetime of these enzymes. Of the five methods known for enzyme immobilization all but encapsulation have been applied to haloperoxidases (Table 7.2). Several of these immobilized enzymes can be purchased (Table 7.3), while others can be readily prepared (Tables 7.4 through 7.6). Some of the observed benefits from immobilization include extended stability toward temperature and organic solvent denaturation (Epton *et al.* 1979, Thibault *et al.* 1981, Darkanbaev *et al.* 1984), and extended halogenation activity (Neidleman 1981, Tenovuo & Kurkijarvi 1981).

Table 7.2 – Immobilization techniques used with haloperoxidases.

Technique	Examples	Reference
Adsorption on solid supports	LPO on glass beads	Honka *et al.* (1982)
	CPO on DEAE-Sephadex A-50 beads	Neidleman *et al.* (1981)
	LPO on hydroxyapatite	Tenovuo *et al.* (1977)
	LPO on zirconium phosphate	Marinchenko (1981)
	HRPO on chromic oxide	Reed (1980)
	LPO on phenoxyacetyl-cellulose	Bakker & Kaspersen (1979)
Covalent attachment to solid supports	HRPO on arylamino-derivatized porous glass	Olsson & Ogren (1983)
	LPO coupled to polystyrene beads	Miller *et al.* (1976)
Entrapment in polymeric gels	HRPO in Biogel P-300	Sergeev *et al.* (1981)
	LPO in Affi-Gel 401	Pommier & Cahnmann (1979)
	HRPO in gelatin	Renneberg *et al.* (1982)
Intermolecular cross-linking	LPO in a copolymer of maleic anhydride and butanediol divinylether	Karonen (1981)

Table 7.3 – Commercially available immobilized haloperoxidases.

Lactoperoxidase on Sephadex beads
Sigma Chemical Co., St. Louis, MO 63172
Lactoperoxidase/glucose oxidase on enzymobeads
Bio-Rad, Richmond, CA 94804

Table 7.4 – Immobilization of chloroperoxidase by adsorption on DEAE-Sephadex A-50 beads.

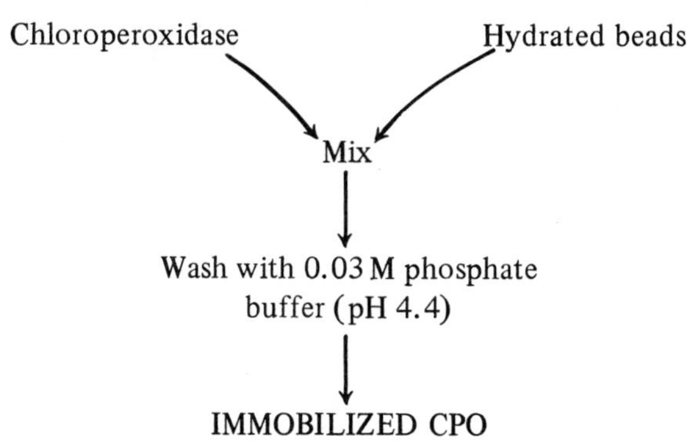

Reference: Neidleman *et al.* (1981).

Table 7.5 – Immobilization of horseradish peroxidase by covalent attachment to porous glass.

3-Aminopropyltriethoxysilane (10% solution) + Porous glass (acid washed)

↓

Prepare aminated glass
(3 hr, 75°C, pH 3.45)

↓ [3% p-Nitrobenzoychloride in $CHCl_3$ with 5% triethylamine]

Acylate the aminated glass
(4hr, reflux)

↓ [5% sodium dithionite]

Reduce the nitro group
(1 hr, boiling)

↓

Dry overnight
(95°C)

↓ [2M HCl plus sodium nitrite]

Diazotize the glass
(30 min, 0°C (50 mg)

Horseradish peroxidase (5 mg) + Diazotize the glass

↓

Incubate
(overnight, 4°C, pH 8.5)

↓

Wash on a glass filter

↓

IMMOBILIZED HRPO

Preparations stored for 6 months at 4°C do not lose significant activity.
Reference: Olsson & Ogren (1983).

Table 7.6 – Immobilization of lactoperoxidase by intermolecular cross-linking in a copolymer of maleic anhydride and butanediol divinylether.

0.5 M Sodium phosphate buffer
(7 ml)

Lactoperoxidase (15 mg) ↘ ↓ ↙ Copolymer (100mg)

↓

Incubate and mix solution in
a pH-Stat
(pH 7.0–7.5, 0°C, 60 min)

↓

Wash and centrifuge, 3X

↓

Homogenize
(0°C)

↓

Wash and centrifuge, 3X

↓

Resuspend in a minimal amount
of sodium acetate buffer

↓

IMMOBILIZED LPO

Reference: Karonen (1981).

THE POSSIBILITIES

The present commercial applications of haloperoxidases are only a tiny fraction of the established biocatalyst market, which, in itself, is only a tiny fraction of the total catalyst market. Therefore, out of necessity, this discussion has to focus primarily on *potential* uses for enzyme-catalyzed halogenation, rather than established uses. From an optimistic viewpoint, one can speculate about a wide range of potential commercial applications:

1. As a synthetic tool

Halogenation is a key step in the production of many industrial chemicals. In some cases, the end product contains halogen; in others, the halogenated chemical is an intermediate (Fig. 7.2). Although haloperoxidase-catalyzed halogenation can yield many of these chemicals, that in itself is not enough incentive to switch to an enzyme-based process. Haloperoxidase-based processes face stiff competition from established and economical chemical halogenation processes. For example, an enzymatic process for converting propylene to propylene oxide via propylene halohydrin has been developed (Fig. 7.3). Haloperoxidase was used to convert propylene to propylene halohydrin and then a second enzyme, halohydrin epoxidase, was used to convert the halohydrin to propylene oxide and halide ion. The halide ion could then be recycled for re-use. Various enzymatic approaches were used to generate the H_2O_2 required by the haloperoxidase. Although technological goals were achieved, the economics of the enzymatic process did not warrant replacement of existing chemical chlorohydrin manufacturing plants. In addition, the desire of the chemical industry to move away from any halogen-based reactions, where possible, because of pollution concerns and costs, did not help. Such is the case in the chemical manufacture of propylene oxide, where the halogen-dependent chlorohydrin route is being rapidly replaced by the halogen-independent Oxirane process. If a haloperoxidase-based process could carry out a halogenation reaction that was difficult or impossible by chemical halogenation, then there might be incentive to use these enzymes. As discussed in Chapter 4, haloperoxidases have not produced optically active products, so they so not possess the advantage of stereospecificity associated with many enzyme-catalyzed reactions. However, there is promise for haloperoxidases in the specialty chemical area. For example, Franssen & van der Plas (1984) compared chlorination of barbituric acid and its derivatives by chloroperoxidase-catalyzed chlorination and by chemical chlorination processes. They noted that the enzymatic system gave a higher yield of a purer product, probably owing to the mild chlorination conditions (Fig. 7.4).

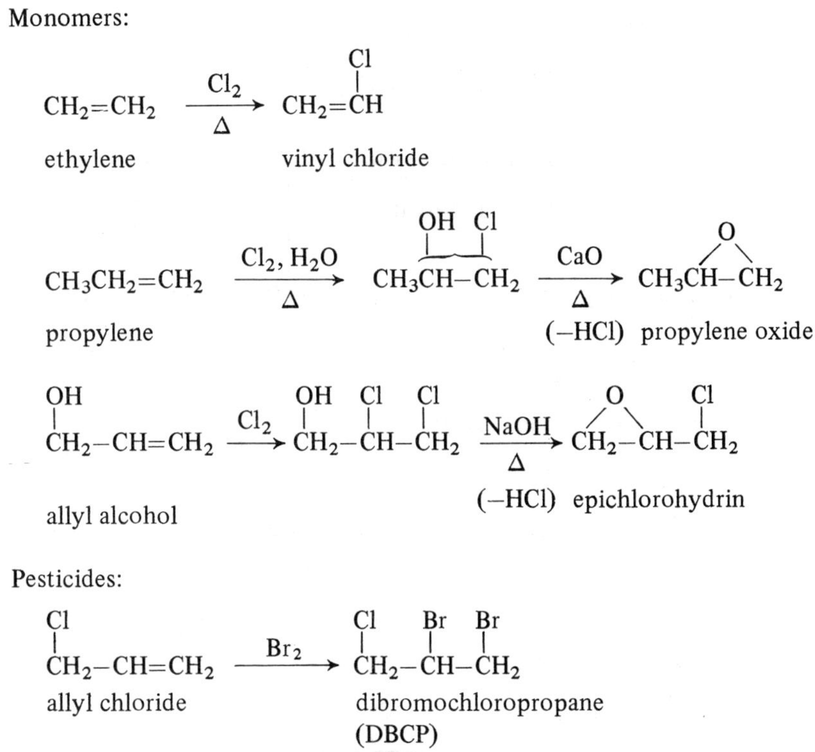

Fig. 7.2 – Halogenation as a key step in the production of some industrial chemicals.

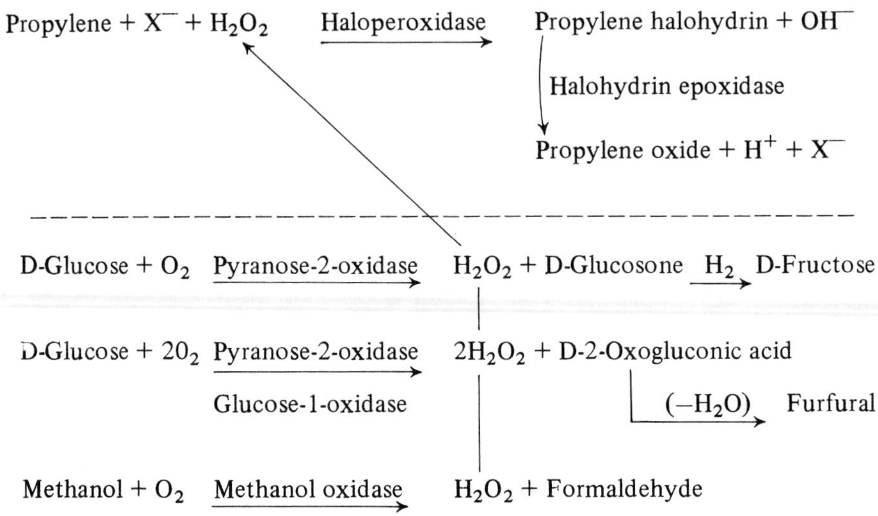

Reference: Neidleman (1980).

Fig. 7.3 – Cetus three-enzyme process for converting propylene to propylene oxide.

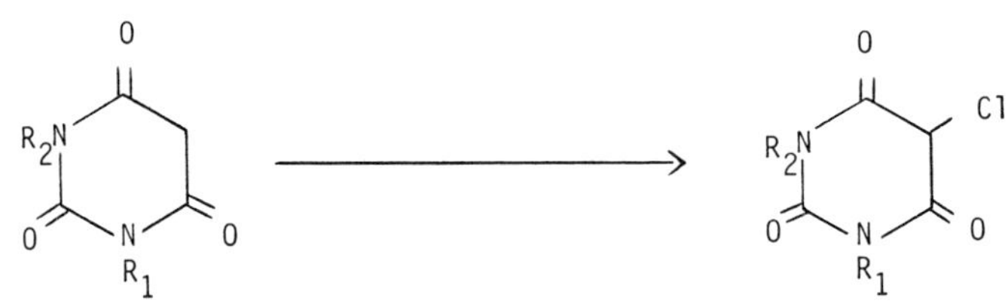

R_1, R_2 = H or Me

Chlorination condition	*Product yield*
$CPO/H_2O_2/Cl^-/H_2O$	>90%
Cl_2/H_2O	~40%
$Cl^-/H_2O_2/\Delta$/dioxane	~80%

Reference: Franssen & van der Plas (1984).

Fig. 7.4 – Comparison of chlorination processes for barbituric acid and its methyl derivatives.

The use of haloperoxidases as peroxidases is already well established in synthetic chemistry. Better product specificity than that normally obtained by chemical peroxidation processes occurs. Such is the case in the syntheses of coumarinolignans by oxidative coupling with horseradish peroxidase (Lin & Cordell 1984), of antihypertensive glycoside derivatives by oxidative dimerization of coniferyl alcohol with chloroperoxidase (Sih 1978), and of nitroso aromatic compounds by oxidation of anilines with chloroperoxidase (Corbett & Corbett 1983).

Klibanov *et al.* (1981) used the oxidase capability of horseradish peroxidase to selectively synthesize the hypotensive agent L-DOPA from L-tyrosine (Fig. 7.5). Hopefully, the continuing examination of haloperoxidases in chemical syntheses will lead to many more additional examples in the near future.

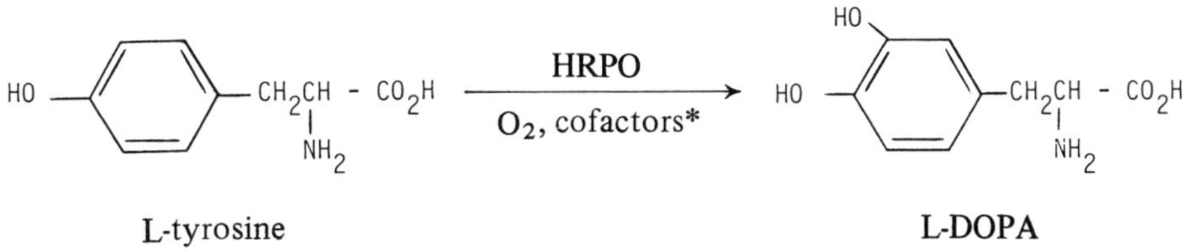

* dihydroxyfumaric acid
Reference: Klibanov *et al.* (1981).

Fig. 7.5 – Horseradish peroxidase-catalyzed synthesis of the hypotensive agent L-DOPA.

2. As a pharmaceutical agent

Recognizing the antibacterial properties of hypohalous acid, haloperoxidases have been used as *in situ* antibacterial-generating agents. For example, lactoperoxidase in combination with thiocyanate ion and urea peroxide, as well as myeloperoxidase in combination with iodide ion, have been formulated with powdered drug preparations to help preserve the drug from bacterial decomposition (Bjorck 1982, Hasegawa & Kobayashi 1983). Based upon the discussion of medical applications for haloperoxidases in Chapter 6, one can envision the possible use of these enzymes as therapeutic agents. If they can be delivered to the site of an infection or tumor (say, for example, by coupling the enzyme to a monoclonal antibody), the haloperoxidases could release their potent oxidant, hypohalous acid, selectively destroying the invading agent or cell. Also, Kessler & Rosenbaum (1984) have obtained a patent on the use of these enzymes as peroxidases in toothpastes, mouthwashes, and chewing gum as a method of rapidly killing oral bacteria to alleviate gingival disease and periodontal disease.

3. In pollution control

Slime growth is a major problem in the process waters of the pulp and paper mill and other industries, resulting in the fouling or plugging of pipes and filters.

To overcome this problem, biocides are added to nonselectively kill a wide spectrum of microorganisms. Hypohalous acid is one such biocide, and a patent has been issued in which a haloperoxidase, halide ion and H_2O_2 are added to the effluent (Orndorff 1983).

However, the most widely proposed use for these enzymes in pollution control is as peroxidases. Horseradish peroxidase is capable of removing carcinogenic aromatic amines from industrial aqueous effluents (Klibanov & Morris 1981), phenols from coal-conversion waste waters (Klibanov *et al.* 1983), low molecular weight color bodies from pulp mill effluents (Plaice & Jurasek 1984), and flavored aromatics from drinking water (Fliessinger *et al.* 1984).

4. For analytical diagnostics

The radioisotopes of iodide, ^{125}I and ^{131}I, and to a lesser extent the radioisotope of bromide, ^{77}Br, are used to label proteins in order to study biological processes. Haloperoxidases provide a means by which the radioisotopes are rapidly, mildly, and stably attached to the protein molecules, yielding high specific radioactivity with retention of the protein's biological activity. Representative examples of these procedures are presented in Tables 7.7 and 7.8.

Table 7.7 – Iodination labeling procedure for thyroid stimulating hormone (TSH) using lactoperoxidase (LPO).

1. Rehydrate the Enymobead reagent* with 0.5 ml distilled water at least 1 hour before use. Store at 4°C.
2. Make up 1% Beta-D-Glucose in aqueous solution. (2% Alpha-D-Glucose can be used; however, it must be allowed to mutarotate overnight to the Beta configuration.)
3. Into a disposable test tube add:

0.2 M phosphate buffer pH 7.2	50 μl
Protein sample	10–25 μl TSH
Enzymobead reagent	50 μl
1.0 mCi $Na^{125}I$	5–25 μl
1% Beta-D-Glucose	25 μl

4. Mix the reagents and allow iodination to proceed at room temperature for 15–25 minutes. The reaction can be quenched by either of two methods. The first requires centrifugation to remove the Enzymobead reagent from the reaction mixture, followed by immediate removal of the supernatant for subsequent gel filtration. The second method utilizes direct application of the test tube mixture to a gel filtration column. In both cases, a Bio-Gel P 100–200 mesh column is recommended for the separation of the unbound iodide from the labeled protein.

* LPO and glucose oxidase co-immobilized on beads.
Reference: Bio-Rad, Technical Bulletin 1071E, Richmond, CA 94804.

Table 7.8 – Bromination labeling procedure for HSA using bromoperoxidase (BPO).

Reaction mixture:	0.8 units*/ml BPO isolated from the alga *Penicillus capitatus* 100 μg/ml of Human Serum Albumin (HSA) 8 μCi/ml of ^{77}Br at approximately 0.08 mM Br^-
Initiation reagent:	80 μM H_2O_2 (final concentration)

Procedure:

1. Add the H_2O_2 to the reaction mixture.
2. Incubate 30 min at 37°C.
3. Add trichloroacetic acid (TCA) to terminate the reaction.
4. Centrifuge and save pellet.
5. Wash pellet with 1 N NaOH and count.

* Units are defined as 1 μmole MCD reacted/min.
Reference: Manthey *et al.* (1984).

Haloperoxidase has also found use as a selective probe for measuring the level of a particular halide ion in a mixture of halide ions. The selective measurement of iodide ion in the presence of bromide, chloride, and fluoride ions has been accomplished by measuring the heat released in the oxidation of iodide ion by horseradish peroxidase (Grime & Lockhart 1979). Such an approach could be applied using other halide ion/haloperoxidase combinations (Table 7.9).

Table 7.9 – Use of haloperoxidases to measure specific halide ion types.

Halide ion type(s) to be measured	Non-interfering halide ion types	Choice of haloperoxidase
I^-	F^-, Cl^- and Br^-	iodoperoxidase
$I^- + Br^-$	F^- and Cl^-	bromoperoxidase
Br^-	F^- and Cl^-	bromoperoxidase
Cl^-	F^-	chloroperoxidase

By far the best known diagnostic uses of these peroxidases are in coupled enzyme assays (for example, to quantitate the amount of glucose in the blood by measuring the amount of H_2O_2 produced in the presence of glucose oxidase) or in enzyme immunoassays (for example, to measure the presence of antigen/antibody complexes in which either species may be labeled with the enzyme).

Various detection schemes for these reactions have been developed (Fig. 7.6). Horseradish peroxidase has been widely used in these assays (Blake and Gould 1984, Murty *et al.* 1984), but there is no reason why other haloperoxidases could not be used, as for example in a recent report in which chloroperoxidase-catalyzed bromination found use in an enzyme immunoassay (Fig. 7.7). Immunoglobulin G (IgG) inhibited chloroperoxidase-catalyzed bromination when the enzyme was conjugated to IgG antibody.

Conversion of nonfluorescent substrate to fluorescent product

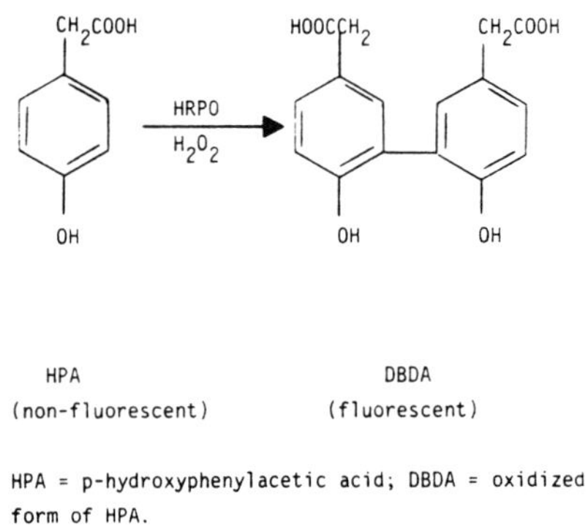

Formation of colored product

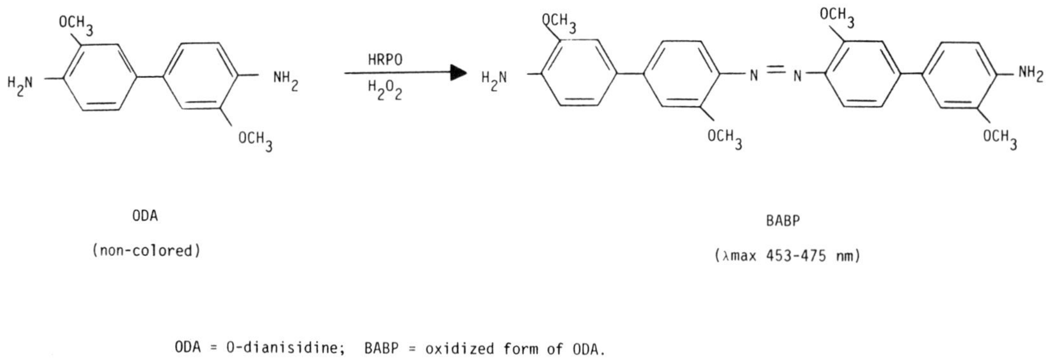

Fig. 7.6 – Detection schemes for assays utilizing peroxidase.

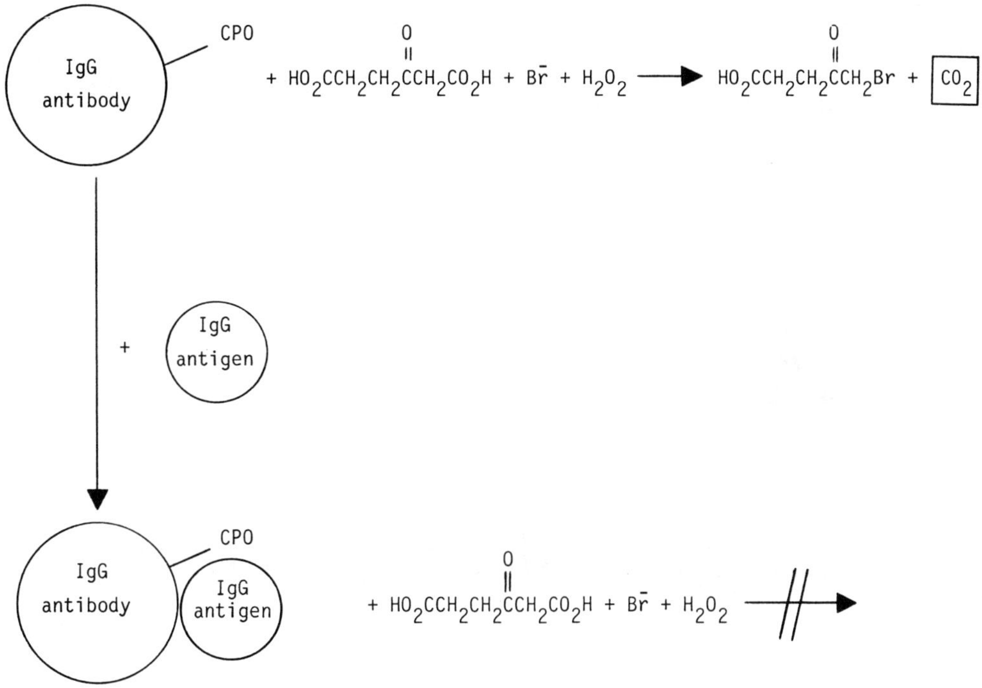

□ = CO_2 released is measured by a CO_2 gas-sensing electrode.
Reference: Fonong & Rechnitz (1984).

Fig. 7.7 – Enzyme immunoassay utilizing chloroperoxidase – catalyzed bromination.

SUMMARY

It is difficult to predict where or when haloperoxidases will achieve significant commercial utility. It is clear that much work is needed to establish desirable characteristics for the enzymatic reaction to satisfy the requirements for a commercial process. It is also clear that a meaningful commercial target has not yet been defined, one, for example, that would not face stiff competition from existing chemical halogenation processes. *For now*, at any rate, haloperoxidases are finding commercial applications not as halogenating agents but as peroxidases.

8

Biological dehalogenation

INTRODUCTION

Before the development of large-scale industrial activities, the concentrations of halo-organic chemicals on the surface of the earth remained reasonably constant with biosynthesis and biodegradation in equilibrium owing to the integrated activities of plants, animals, and, particularly, the microbes. However, because of man's activity, a host of halogenated compounds, many of which are recalcitrant, have been thrust into the natural ecosystem. There are well over 100 different halogenated compounds (fluorine, chlorine, bromine, and iodine containing) in commercial use today, with a total world production exceeding 20 million tons/year. Figs. 8.1 through 8.3 illustrate the diversity of the compounds in use. Many of these halogenated compounds appear in the natural ecosystem either by intent (e.g., the use of halogenated pesticides), or because of poor practices (e.g., the volatilization of the halogenated lead scavengers in gasoline), or through abuse (e.g., the illegal dumping of wastes). On our planet there are now several locales plagued with heavy contamination by halogenated compounds, for example, Agent Orange in the jungles in Southeast Asia, dioxin in Seveso in Italy, and kepone in the James River of Virginia.

Not all halogenated compounds found in the environment are of direct industrial origin. The listing of natural halometabolites presented in Chapter 2 is evidence of this. In addition, as a special case, natural sources such as forest fires and decomposition of seaweeds release up to 100 times more chloromethane

Solvents	*Gasoline additives*	*Fumigants*
CH_3Cl chloromethane	CH_2ClCH_2Cl 1,2-dichloroethane	CH_3CH_2Br bromoethane
CH_2Cl_2 methylene chloride	CH_2BrCH_2Br 1,2-dibromoethane	$CH_2ClCH{=}CHCl$ 1,3-dichloropropene
CH_3CCl_3 1,1,1-trichloroethane		$CH_2ClCHBrCH_2Br$ 1,2-dibromo-3-chloropropane

Fig. 8.1 – Industrial halogenated aliphatics.

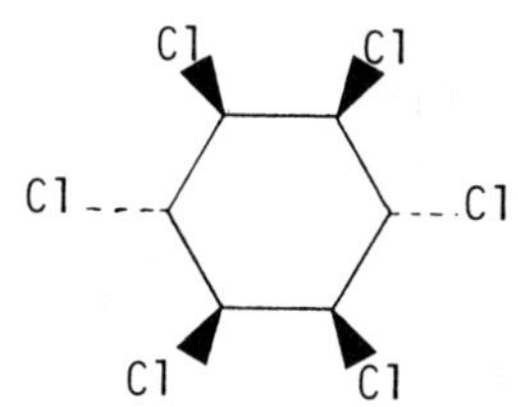

lindane

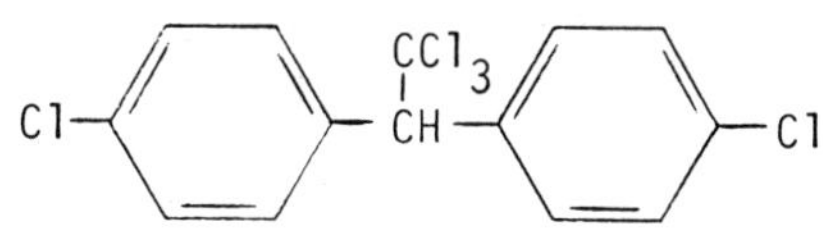

DDT

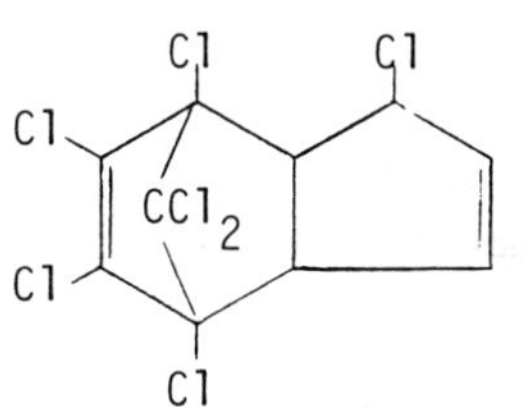

heptachlor

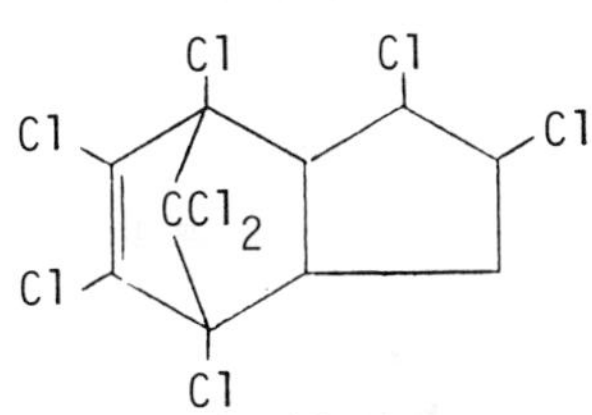

chlordane

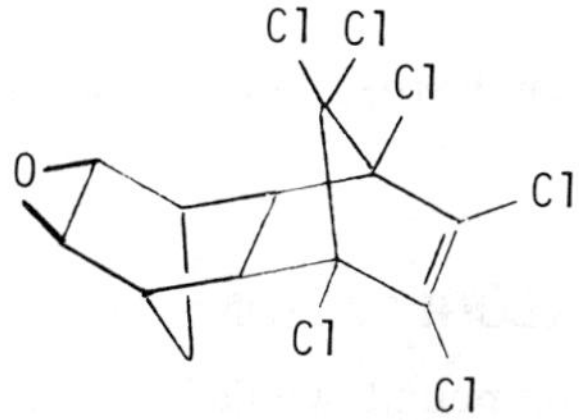

dieldrin

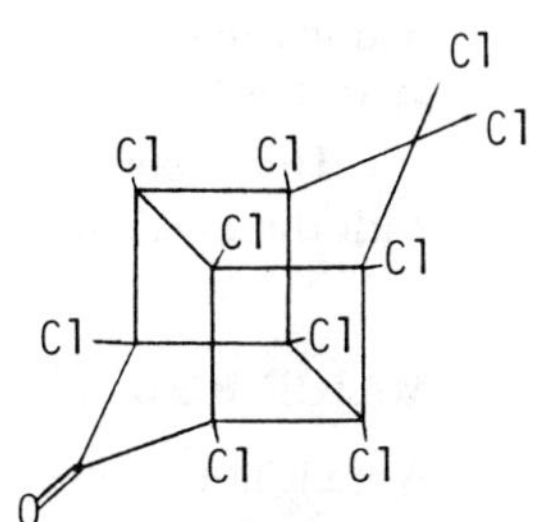

kepone

Fig. 8.2 – Halogenated insecticides.

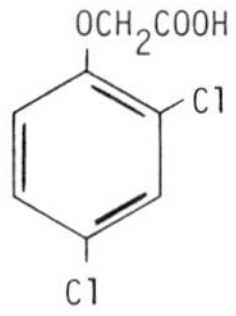

2,4-D

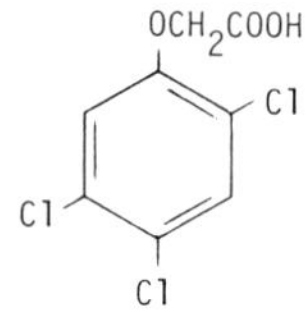

2,4,5-T

(Agent Orange is the butyl esters of 2,4-D and 2,4,5-T, in a 1:1 ratio.)

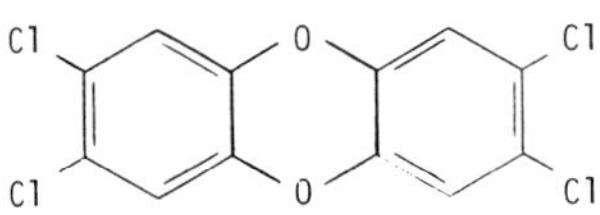

dioxin (TCDD)

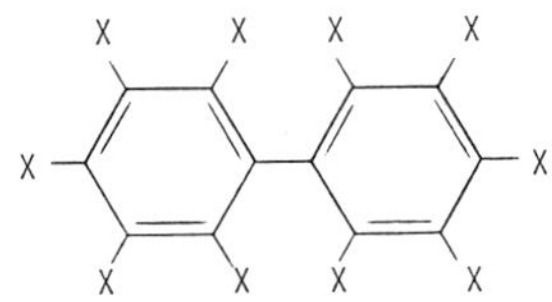

X = H or Cl
PCBs

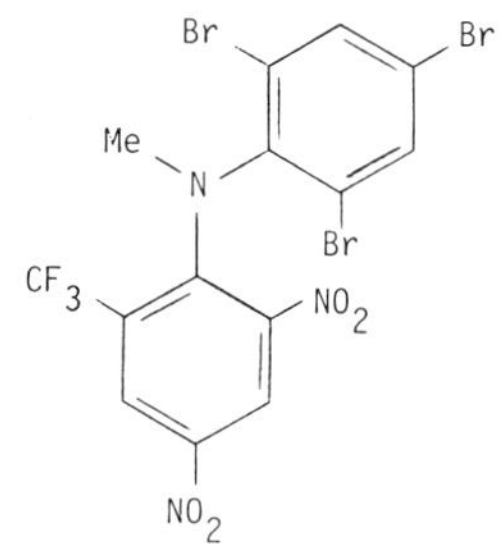

bromethalin
(rodenticide)

FCH_2CO_2Na

sodium fluoroacetate
(coyote poison)

Fig. 8.3 – Other halogenated compounds added to the environment.

than that manufactured by the chemical industry (Leisinger 1983). The macroalgae have also been implicated as an important source of organobromine compounds released into the atmosphere, approximately 10^4 tons/year (Gschwend *et al.* 1985).

This chapter will discuss how nature has learned, and is learning, to deal with these halogenated xenochemicals.

MAJOR ROUTES OF BIOLOGICAL DEHALOGENATION

Were it not for microbes, these halogenated compounds would be rapidly polluting the face of the earth. Many naturally occurring microorganisms have been isolated that degrade halogenated compounds to some degree or another, as illustrated by Table 8.1. For some recalcitrant compounds like PCBs (polychlorinated biphenyls), successful degradation has been accomplished only in the presence of mixed cultures (Hankin & Sawhney 1984). Table 8.2 gives an indication of how difficult it is, even in the presence of this abundant, metabolizing, natural ecosystem, to degrade some of these compounds.

Table 8.1 – Some naturally occurring microorganisms that degrade halogenated compounds.

Halogenated compound	Microorganism
DDT	*Streptomyces aureofaciens* *Klebsiella pneumoniae* *Aerobacter aerogenes* *Escherichia coli* *Norcardia erythropolis* *Mucor alternans* *Proteus vulgaris*
Aldrin	*Aspergillus niger* *Penicillium chrysogenum* *Trichoderma viride* *Pseudomonas* sp. *Micrococcus* sp.
Lindane	*Bacillus coli* *Clostridium sphenoides* *Chlorella vulgaris* *Pseudomonas putida*
2,4-D	*Achromobacter* sp.
Dalapon	*Pseudomonas putida*
Silvex	*Pseudomonas* sp. *Achromobacter* sp.

Reference: Lal & Saxena (1982), Johnston & Robinson (1983).

Table 8.2 – Persistence of some halogenated pesticides.

Pesticide	Period of time at which the pesticide was still detectable	Approximate half-life
Chlorine	21 yr	2–4 yr
DDT	24 yr	3–10 yr
Dieldrin	21 yr	1–7 yr
Heptachlor	16 yr	7–12 yr
Toxaphene	16 yr	10 yr

Reference: Alexander (1977).

Five major pathways for enzymatic degradation of halogenated compounds have been discovered. Each of these pathways will be discussed briefly:

1. Reductive dehalogenation

The reaction proceeds by replacing the halogen atom with a hydrogen atom:

$$C{-}X \longrightarrow C{-}H + X^-$$

Such a mechanism has been invoked for the initial step in the anaerobic degradation of DDT (Fig. 8.4), the metabolism of chloropicrin by cytochrome P-450 (Fig. 8.5), and surprisingly, the anaerobic degradation of haloaromatic substrates (Fig. 8.6). In this latter example, the primary degradation event was loss of the aryl halogen without alteration of the aromatic ring.

A membrane-bound reductive dehalogenating enzyme, thyroid hormone 5′-deiodinase, has been found in lysosomes prepared from rat liver and kidney (Colquhoun & Thomson 1985). Another reductive dehalogenating enzyme has even been isolated: iodotyrosine deiodinase from the thyroid (Fig. 8.7). This enzyme probably plays a role in glandular iodine conversion.

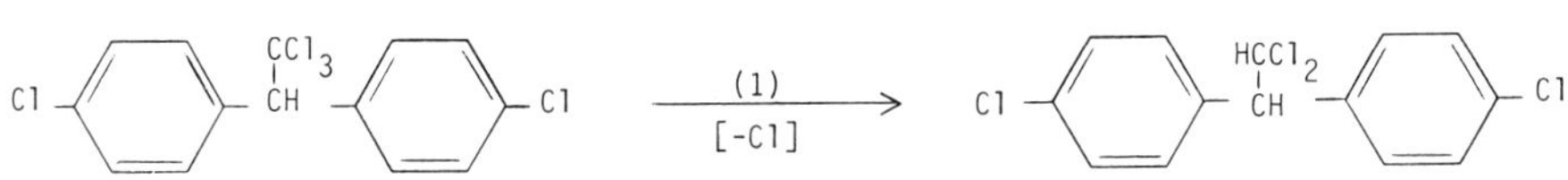

DDT DDE

(1) Actinomycetes: *Norcardia* sp.
Streptomyces aureofaciens
S. cinnamoneus
S. viridochromogenes

Reference: Chacko *et al.* (1966).

Fig. 8.4 – Reductive dehalogenation of DDT.

$$\underset{\text{chloropicrin}}{Cl_3CNO_2} \xrightarrow[(-Cl)]{} Cl_2CHNO_2 \xrightarrow[(-Cl)]{} ClCH_2NO_2 \xrightarrow[(-Cl)]{} \underset{\text{nitromethane}}{CH_3NO_2}$$

Reference: Castro *et al.* (1985).

Fig. 8.5 – Three successive reductive dehalogenations of chloropicrin by cytochrome P-450 of *Pseudomonas putida*.

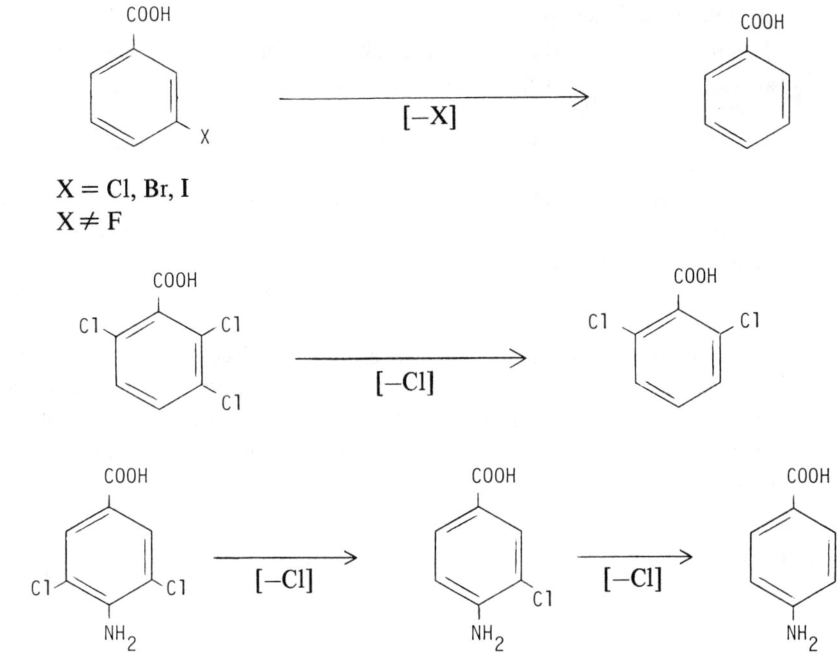

References: Suflita *et al.* (1982), Horowitz *et al.* (1983).

Fig. 8.6 – Reductive dehalogenation of halobenzoates by anaerobic microbial communities.

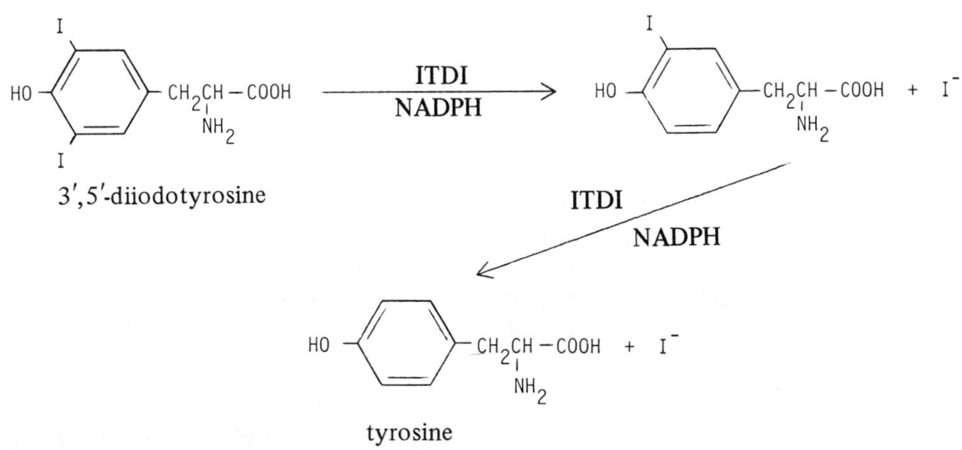

Reference: Rosenberg & Goswami (1984).

Fig. 8.7 – Iodotyrosine deiodinase (ITDI), a reductive dehalogenating enzyme.

2. Dehydrohalogenation

The reaction proceeds by the simultaneous removal of the halogen atom and the adjacent hydrogen on the neighboring carbon:

$$\text{H–C–C–X} \longrightarrow \text{C=C} + \text{HX}$$

Such a mechanism has been proposed for the initial step in the aerobic degradation of DDT (Fig. 8.8), and the degradation of lindane (Fig. 8.9). Simple halogenated hydrocarbons are also converted to alkenes by dehydrohalogenation. For example, ethylene dibromide is quantitatively converted to ethylene by soil microbes (Castro 1977).

Dehydrohalogenation is not always due to direct enzymatic dehalogenation, but instead can be due to spontaneous decomposition of an enzyme-produced intermediate. Such is the case with the enzymatic hydration of fluorofumarate by fumarase and the enzymatic oxidation of fluoromethylglyoxal by glyoxalase (Fig. 8.10).

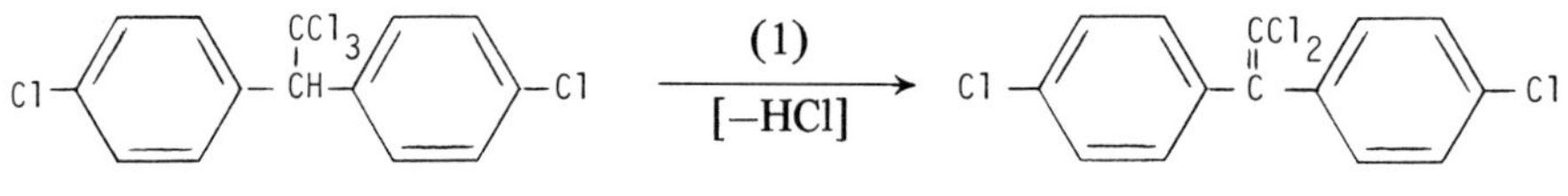

DDT DDE

(1) Algae: *Skeletonema costatum*
Dunaliella sp.
Cyclotella nana
Isochrysis galbana
Insects: *Culex pipiens* var. molestus, ovarian tissue
House fly, DDT-resistant strain SPK52 DDT24 R4

References: Lal & Saxena (1982), Clark & Shamaan (1984), Yoshida *et al.* (1985).

Fig. 8.8 – Dehydrohalogenation of DDT.

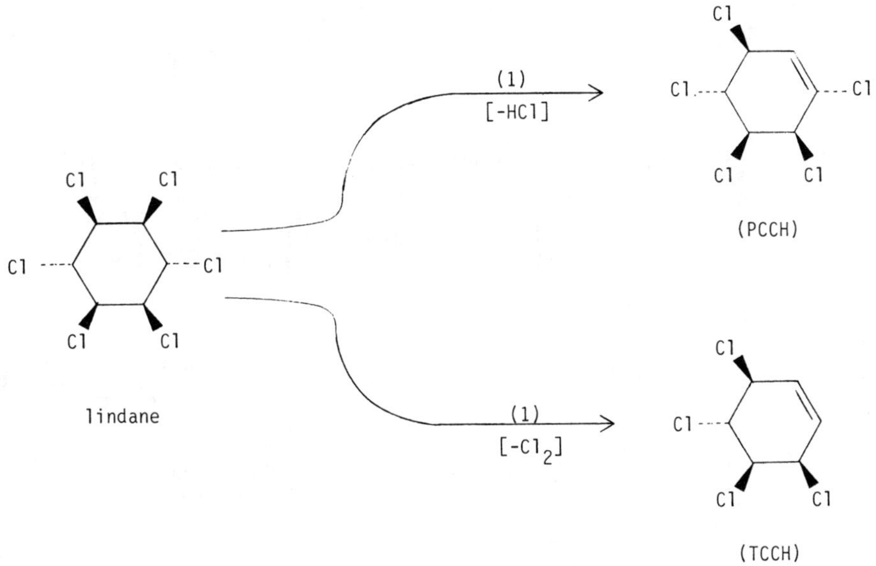

(1) Dehydrochlorination is carried out by numerous microorganisms, some of which yield only PCCH (e.g., *Escherichia coli*), some yield only TCCH (e.g., *Clostridium rectum*), and some yield both PCCH and TCCH (e.g., *Pseudomonas putida*).

Reference: Lal & Saxena (1982).

Fig. 8.9 – Dehydrohalogenation of lindane.

A. Hydration of fluorofumarate by fumarase (Marletta *et al.* 1982).

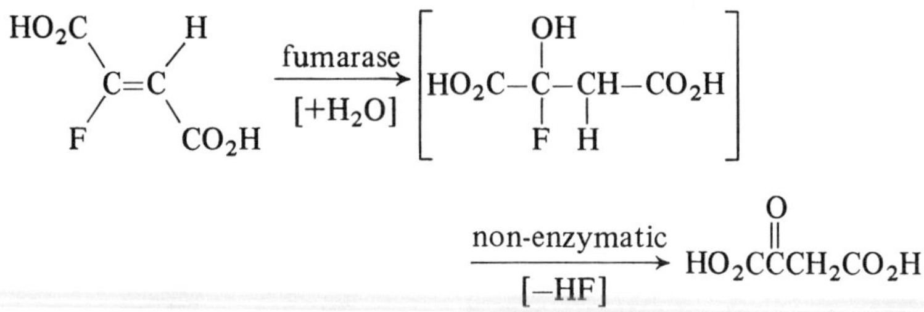

B. Oxidation of fluoromethylglyoxal by glyoxalase (Kozarich *et al.* 1981).

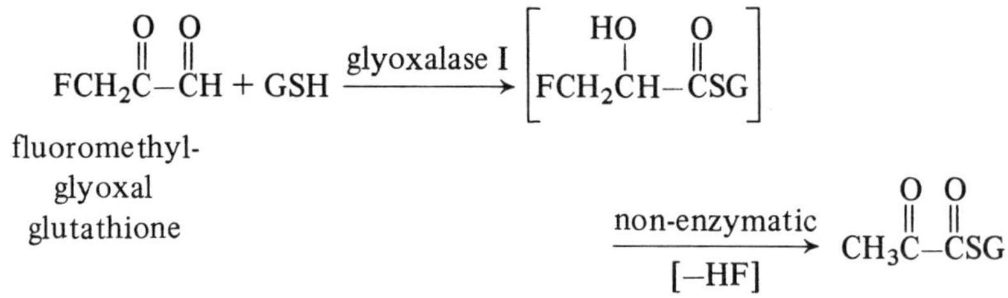

Fig. 8.10 – Nonenzymatic dehydrohalogenations as a result of a preceding enzymatic step.

3. Hydrolytic dehalogenation

The reaction proceeds by nucleophilic substitution replacing the halogen atom with a hydroxyl ion:

$$\text{C–X} + \text{OH}^- \longrightarrow \text{C–OH} + \text{X}^-$$

Purified enzymes have been obtained that utilize this dehalogenating pathway. These enzymes, called halidohydrolases, are divided into two groups: haloacetate dehalogenase (EC 3.8.1.3), which acts exclusively on haloacetates, and 2-haloacid dehalogenase (EC 3.8.1.2), which acts on haloacetates and other short-chain fatty acids. Representatives of these two enzyme groups are presented in Tables 8.3 and 8.4. The facile defluorinating ability of some of the haloacetate dehalogenases is most impressive when compared to the refractory release of fluoride ion from fluoroacetate even in the presence of boiling 10% NaOH. Even though the C–F bond is the strongest of the carbon–halogen bonds (107, 67, 54, 46 kcal/mol; C–F, C–Cl, C–Br, C–I; bond energies, respectively), these enzymes preferentially defluorinate rather than dechlorine, debrominate, or deiodinate. Such defluorinases have been discovered not only in microbes but also in the livers of various mammals (Soiefer & Kostyniak 1984). Studies have shown that the enzymatic elimination of halogen atoms proceeds through an S_N2-type reaction, following an inversion of configuration, i.e., L-isomer is converted to D-isomer. A proposed mechanism for this enzymatic dehalogenation is shown in Fig. 8.11.

Table 8.3 – Substrate specificity of some haloacetate dehalogenases.

Substrate	X	*Pseudomonas* sp. (1)	*Moraxella* sp. (2)	*Fusarium solani* (3)
XCH_2CO_2H	F	100	0	100
	Cl	21	63	25
	Br	14	100	0
	I	0	94	0
X_2CHCO_2H	Cl	0	2	–
X_3CCO_2H	Cl	0	0	–
CH_3CHXCO_2H	Cl	0	6	–
$XCH_2CH_2CO_2H$	Cl	0	0	–
$CH_3CX_2CO_2H$	Cl	0	0	–
$CH_3CH_2CHXCO_2H$	Cl	0	0	–

References:
(1) Kawasaki *et al.* (1981a).
(2) Kawasaki *et al.* (1981b).
(3) Walker & Lien (1981).

Table 8.4 – Substrate specificity of some 2-haloacid dehalogenases.

Substrate	X	*Pseudomonas* sp. (1)	*Pseudomonas putida* (2)	*Rhizobium* sp. 1 (3)	*Rhizobium* sp. 2 (3)
XCH_2CO_2H	F	0	0	–	–
	Cl	9	52	0	12
	Br	74	100	100	100
	I	25	71	0	25
X_2CHCO_2H	Cl	2	8	13	2
X_3CCO_2H	Cl	1	0	0	1
CH_3CHXCO_2H					
D	Cl	22	0	–	–
L	Cl	31	28	–	–
D, L	Cl	26	–	22	9
D, L	Br	100	69	26	47
$XCH_2CH_2CO_2H$	Cl	0	0	–	–
$CH_3CX_2CO_2H$	Cl	11	3	12	4
$CH_3CH_2CHXCO_2H$	Cl	4	0	8	4
	Br	58	27	–	–

References:
(1) Motosugi *et al.* (1982a).
(2) Motosugi *et al.* (1982b).
(3) Allison *et al.* (1983).

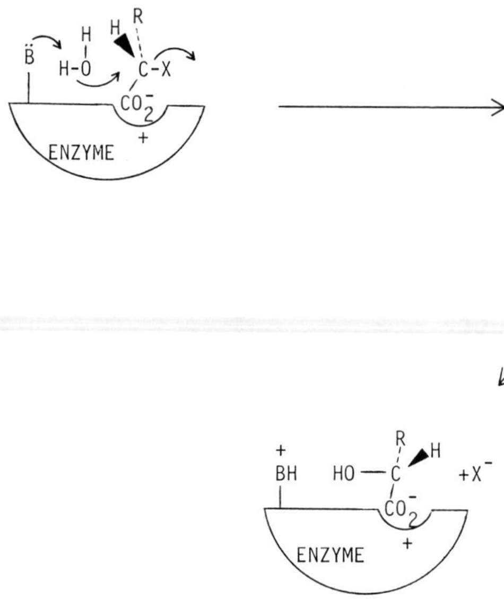

Reference: Motosugi & Soda (1983).

Fig. 8.11 – Proposed reaction mechanism for halidohydrolases.

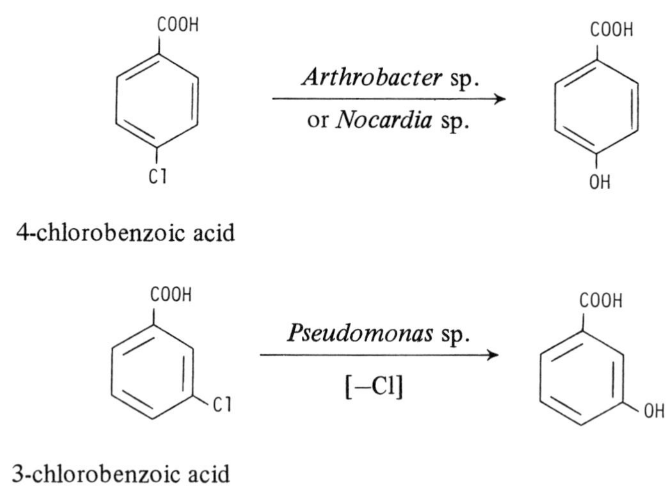

References: Johnston *et al.* (1972), Klages & Lingens (1979), Marks *et al.* (1984).

Fig. 8.12 – Proposed pathway for hydrolytic dehalogenation of aromatics.

Hydrolytic dehalogenation has been proposed as a mechanism for degradation of short-chain halogenated hydrocarbons such as dichloromethane and dichloroethane. The resulting short-chain alcohols can then be further metabolized by oxidation with alcohol dehydrogenase to yield readily assimilated aldehydes like formaldehyde (Janssen *et al.* 1985, Kohler-Staub & Leisinger 1985).

Hydrolytic dehalogenation of halogenated benzoates also occurs. As with reductive dehalogenation, this pathway of enzymatic dehalogenation removes halogen without alteration of the aromatic ring (Fig. 8.12).

This biochemical nucleophilic substitution can proceed with nucleophiles other than hydroxyl ion, such as thiols like glutathione.

4. Epoxidation.

The reaction proceeds by the simultaneous removal of the halogen atom and the hydrogen from the adjacent hydroxyl group on the neighboring carbon:

$$\mathrm{HO{-}C{-}C{-}X} \longrightarrow \mathrm{C{-}C} + \mathrm{HX}$$

To date, only a single source of an enzyme employing this dehalogenating pathway has been found. The enzyme, called halohydrin epoxidase, has been found in the bacterium *Flavobacterium* sp. (Castro & Bartnicki 1968).

Alpha, beta-iodohydrins, bromohydrins, and chlorohydrins are suitable substrates for this enzyme (Table 8.5). Comparing a halohydrin series (e.g., the 2-haloethanol series), the enzyme reacts most readily with the bromohydrin. Comparing halohydrin positional isomers (e.g., the two bromopropanol isomers), the enzyme catalyzes the conversion of the terminal halo-containing isomer more effectively. Alpha, gamma-halohydrins (e.g., $BrCH_2CH_2CH_2OH$) and alpha, beta-fluorohydrins (e.g., FCH_2CH_2OH) are not substrates for this enzyme. This order of reactivity is identical with that observed in the chemical reaction in which halohydrins are treated with bases such as lime.

Because the enzymatic reaction is reversible, transhalogenation (i.e., exchange of halogen atoms) can occur with alpha-halo epoxides (Bartnicki & Castro 1969):

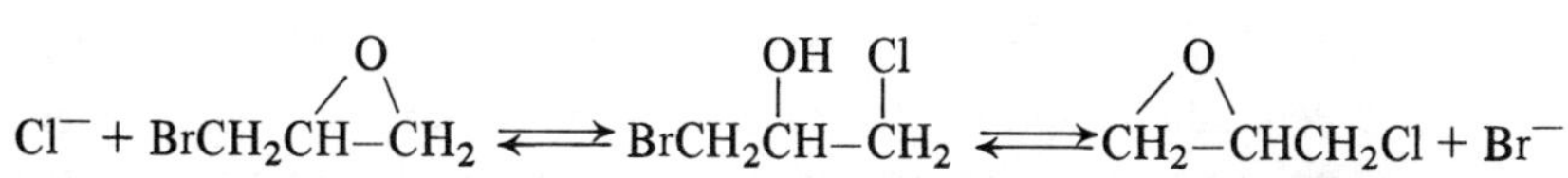

epibromohydrin epichlorohydrin

As discussed in Chapter 7, enzymatic dehalogenation of halohydrins has been coupled to another enzymatic reaction – the halogenation of alkenes by haloperoxidases – as a two-step enzymatic conversion of alkenes to epoxides.

Table 8.5 – Relative activity of whole cells of *Flavobacterium* sp. on the enzymatic synthesis of epoxides from alpha, beta-halohydrins.

$$\underset{\text{RCH}-\text{CHR}'}{\overset{\text{OH}\quad\text{X}}{|\qquad|}} \xrightarrow{\text{Halohydrin epoxidase}} \text{RCH}\overset{\text{O}}{\diagup\diagdown}\text{CHR}' + \text{H}^+ + \text{X}^-$$

Substrate	R	R′	X	Relative activity*
2-fluoroethanol	H	H	F	0
2-chloroethanol	H	H	Cl	18
2-bromoethanol	H	H	Br	60
2-iodoethanol	H	H	I	24
1-chloro-2-propanol	CH_3	H	Cl	26
1-bromo-2-propanol	CH_3	H	Br	100
2-bromo-1-propanol	H	CH_3	Br	5
1-iodo-2-propanol	CH_3	H	I	34
1-bromo-3-butene-2-ol	$CH_2{=}CH$	H	Br	65

* Relative to activity on 1-bromo-2-propanol (taken as 100%) at 7 hr into the reaction.

Reference: Geigert *et al.* (1983g).

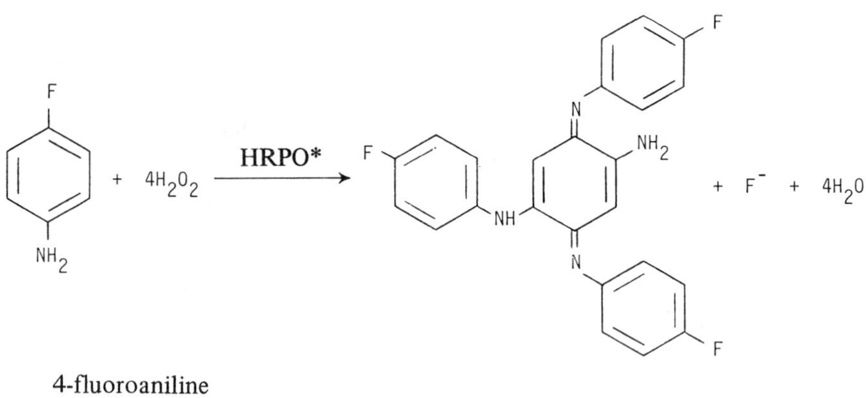

4-fluoroaniline

* Peroxidase rupture of carbon–halogen bonds also occurs when the halogen is chlorine, bromine, or iodine, but the exact structures of the products are uncertain. Other haloperoxidases carry out the same reaction on other halo-aromatics.

Reference: Saunders (1957), Van Zyl *et al.* (1981), Siddiqi (1982).

Fig. 8.13 – Oxidative displacement of halogen via peroxidase.

5. Oxidative displacement

The rupture of a covalent carbon-halogen bond proceeds through polymerization of free radical intermediates or oxidative hydroxylation adjacent to the halogen. The former pathway occurs with peroxidases (Fig. 8.13), while the latter pathway is a characteristic of oxygenases (Fig. 8.14). Both peroxidases and oxygenases are broadly distributed in nature.

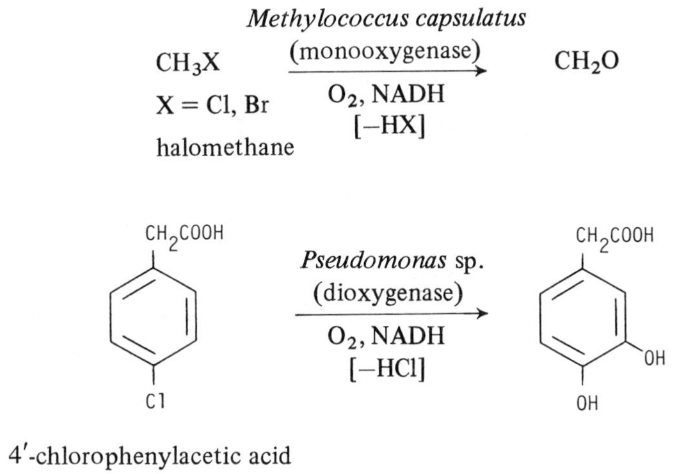

References: Dalton (1981), Markus *et al.* (1984).

Fig. 8.14 – Oxidative displacement of halogen via oxygenases.

INDUCED DEHALOGENATION BY GENETIC ENGINEERING

A variety of genetic engineering techniques have been applied to the dehalogenating enzyme systems. Plasmids have been shown to play a major role in the biodegradation of halogenated compounds. A list of some of the known plasmids encoding for degradation of halogenated compounds is given in Table 8.6. Since plasmids are transmissible by conjugation, this dehalogenating capability can be transferred among the pseudomonads and even to *E. coli* and other Gram-negative bacteria. Thus, by means of continuous culture in a chemostat, it has been possible to breed a microorganism that can carry out a specific dehalogenation process. For example, *Pseudomonas putida* PP3 (a bacterium that can dehalogenate 2-chlorobutanoate to 2-hydroxybutanoate) and HB2001 (a bacterium that grows on 2-hydroxybutanoate) were inoculated into a chemostat. Recombinants were formed that could both degrade 2-chlorobutanoate and grow on its dehalogenation product (Hass 1983). Another example of a bred recombinant is the *Pseudomonas cepacia* strain obtained from a chemostat into

Table 8.6 – Some plasmids encoding for degradation of halogenated compounds.

Plasmid	Degradation pathway
pAC25	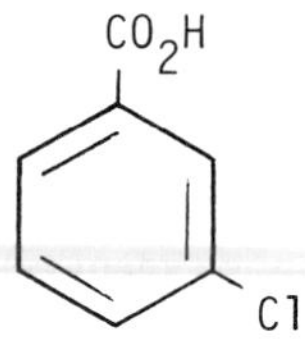 3-chlorobenzoic acid → $HO_2CCH_2CH_2CO_2H$ + CH_3COCoA
pJP1	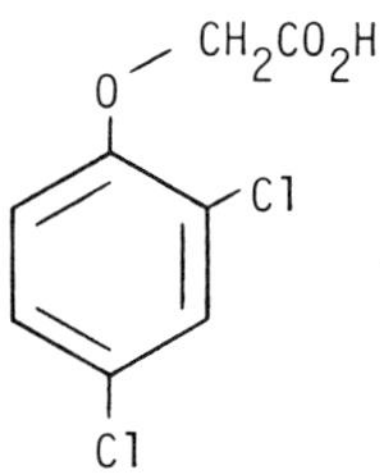 (2,4-D) → $HO_2CCH_2CH_2CO_2H$ + CH_3COCoA + HO_2CCHO
pKF1	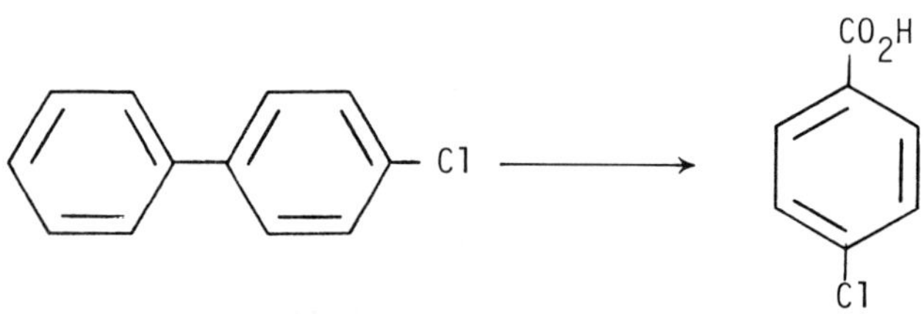 4-chlorobiphenyl
pUO1	$FCH_2CO_2H \longrightarrow HOCH_2CO_2H$ fluoroacetic acid
pUU204	$CH_3CH(Cl)CO_2H \longrightarrow CH_3COCO_2H$ 2-chloropropionic acid

Reference: Haas (1983), Karns *et al.* (1984), Ghosal *et al.* (1985).

which soil samples from a number of chemical waste dump sites and *P. putida* strains harboring a variety of plasmids encoding for dehalogenation were added. By increasing the concentration of 2,4,5-trichlorophenoxyacetate (2,4,5-T) in the chemostat over a seven-month period, a strain of *P. cepacia* called AC1100 was isolated that could utilize 2,4,5-T as a sole source of carbon and energy. Since 2,4,5-T is the refractory component of the defoliant Agent Orange, this accomplishment illustrates the potential of this approach (Chatterjee *et al.* 1980; Karns *et al.* 1984). It has been reported that this *in vivo* genetic selection is being used to breed a recombinant microorganism that can degrade dioxin (*Biotechnology Newswatch*, May 2, 1983, p.1).

A more elegant and direct selection technique is to actually clone the gene(s) coding for the desired enzyme(s). As an example, the DNA fragments coding for three catabolic enzymes (toluate 1,2-dioxygenase, dihydrodihydroxybenzoic acid dehydrogenase, and salicylate hydroxylase) were cloned into pBR322 vector plasmids. The pBR322-based plasmids were then ligated to the cloning vector pKT231, and finally the hybrid plasmids were introduced into *Pseudomonas* sp. B13. This imparted to the *Pseudomonas* sp. B13 derivative the capability of degrading 3- and 4-chlorobenzoate, 3,5-dichlorobenzoate, salicylate and 3-, 4- and 5-chlorosalicylate. The original strain could only degrade 3-chlorobenzoate (Lehrbach *et al.* 1984). Genetic engineering is also being applied in attempts to clone the genes for degrading 2,4,5-T and dioxin (*Biotechnology Newswatch*, August 15, 1983, p.1; New *Scientist*, May 5, 1983, p. 288). Hoechst AG has been reported to have chemically synthesized the DDT-degrading gene and already cloned it into *E. coli* (*Biotechnology Newswatch*, July 16, 1984, p. 7).

There is no doubt that genetically-induced dehalogenation methods can be used to improve the degradative abilities of microbes. Whether or not these engineered strains can be effectively used to control environmental halogen pollution will depend primarily upon how successfully such microorganisms can maintain their new capabilities in nature. It is impossible to predict the stability of the genetically engineered strains in an open system where it is difficult, at best, to maintain some kind of selective pressure.

Further, the present philosophical and regulatory concerns about the dangers of releasing recombinant organisms into the environment have to be overcome. Some positive steps are being taken. The Environmental Protection Agency (EPA) has approved the use of a mixed culture of naturally occurring *Pseudomonas* and *Bacillus* microorganisms for degrading PCB-contaminated sludge (*Chemical Week*, August 29, 1984, p. 44). The city of Seattle, Washington, is considering the use of genetically modified bacteria to decontaminate nearly one million gallons of bunker fuel oil tainted with PCBs (*Chemical Week*, September 19, 1984, p. 28). The city of Madison, Wisonsin, is funding research on biodegrading PCBs in sewage sludge (*Biotechnology Newswatch*, March 4, 1985, p. 3).

Finally, the question should be posed: does enzymatic dehalogenation play a role in biosynthetic pathways in nature? The answer lies in future studies.

9

Future directions for biological halogenation

INTRODUCTION

Previous chapters have described selected information on biological halogenation from diverse segments of the scientific literature: the enzymological, the biochemical, the microbiological, the botanical, the marine, the medical, and the industrial. It is evident that while much is known, much more is unknown. The basic roles of the halogenating enzymes and the halometabolites in nature and medicine will be defined during the course of future research, as will their potential for commercial applications. The brief discussion that follows will survey some areas for future study.

HALOGENATING ENZYMES

The broad specificity of halogenating enzymes dictates that new *in vitro* and *in vivo* substrates will be identified. These reactions, in conjunction with those already known, will help to further clarify the functions of the enzymes in nature, especially in the marine environment and in mammalian medical biochemistry.

In the medical area, there will be a continuing and growing translation of *in vitro* observations into *in vivo* conclusions, leading to novel medical insights and treatments. In particular, controlling the activity of myeloperoxidase (MPO) with specific drugs and directing the activity of MPO to specific targets by binding the enzyme to monoclonal antibodies will be part of future medicine.

The role of the halogenating enzymes in the marine environment will be further clarified. Do these enzymes play a more direct role, for example, in the defense mechanism of algae as in the mammalian polymorphonuclear leukocytes (PMN)? Evidence for the *in vivo* involvement of halogenating enzymes in biosynthetic pathways of marine organisms will be discovered. Do these enzymes play a role in mammalian biochemistry? Are the PMN and eosinophils the sites of unrecognized halide-dependent biosynthetic reactions catalyzed by halogenating enzymes?

Progress will be made in extending our understanding of both the diverse structures of the family of halogenating enzymes and the mechanism of enzymatic halogenation. Since it has been shown that not all halogenating enzymes possess a heme moiety, there is clearly considerable structural divergence in enzymes catalyzing the same reaction – halide activation. In the face of such diverse structural properties, the question of whether the mechanism of enzymatic activity is universely through hypohalous acids or activated halide–enzyme complexes is still open. De Boer et al. (1985) are presently investigating the possibility of stereospecific halogenation with a non-heme, algal bromoperoxidase.

That new halogenating enzymes will be discovered is a foregone conclusion. When the mechanism by which plants carry out fluorination reactions is clarified, a novel enzyme may be uncovered. However, it may be that an enzyme so mundane as horseradish peroxidase, or a related iodoperoxidase, could be responsible for the fluorinated plant metabolites listed in Fig. 2.8, Chapter 2. The enzymatic synthesis of fluoroiodo-derivatives by this enzyme has been reported (Neidleman & Geigert, 1983a). A speculative pathway to obtain the critical building block intermediate, fluoroacetic acid, is shown in Fig. 9.1. The process begins with a terminally unsaturated fatty acid since the reported plant metabolites include terminally fluorinated fatty acids. Having obtained fluoroacetic acid, the biosynthesis of fluorocitrate and fluoroacetone follow from classical biochemistry, assuming some level of enzymatic resistance to fluorometabolites. The soil in which these plants grow should be studied for iodide content as should the plant tissues themselves.

It should be mentioned in passing that a chemical synthesis of the fluorinated nucleoside, nucleocidin, produced by *Streptomyces calvus* (see Fig. 2.8, Chapter 2), involved the use of iodine fluoride (IF) to add the halogens across a double bond, followed by chemical deiodination (Jenkins *et al.* 1976).

With the discovery of new enzymes, new reactions, and improved reaction conditions, the potential for commercial application of these enzymes will be enhanced. Additionally, as the subtleties of variant halogenating enzymes are revealed, the information should have value as a model for the construction of useful catalysts for chemical halogenation.

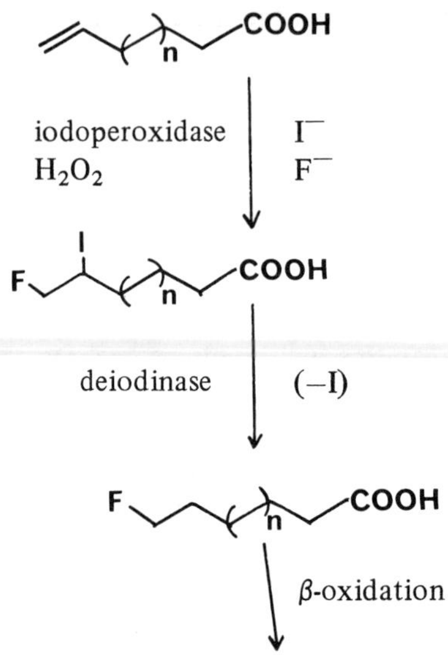

FCH_2COOH	+	CH_3COOH	(even C number fatty acids)
fluoroacetic acid		acetic acid	
FCH_2CH_2COOH	+	CH_3COOH	(odd C number fatty acids)
2-fluoropropionic acid			

$$FCH_2COOH + CH_3COOH \xrightarrow[\text{enzymes}]{\text{several}} FCH_2\underset{\underset{O}{\|}}{C}CH_2COOH$$

3-fluoroacetoacetic acid

$$FCH_2\underset{\underset{O}{\|}}{C}CH_2COOH \xrightarrow[\text{carboxylyase (EC 4.1.1.4)}]{\text{acetoacetate}} FCH_2\underset{\underset{O}{\|}}{C}CH_3 + CO_2$$

fluoroacetone

$$FCH_2COOH \xrightarrow[\text{cycle}]{\text{citric acid}} \text{fluorocitric acid}$$

Fig. 9.1 – Speculative pathways to fluorinated plant metabolites.

HALOMETABOLITES

The number of halometabolites isolated and structurally defined will continue to increase rapidly. The potential usefulness of these compounds for human therapy and, perhaps, for marine management will be driving forces for the continuing pursuit of these metabolites. However, the desire to understand the role of these halometabolites in the lifestyle of marine organisms will also spur on further research. Will some of these compounds be demonstrated to have regulatory or informational value to the producing organism functioning as hormones, attractants, or agents of identification or communication? As indicated in Chapter 5, some of these compounds act as deterrents: anti-feedants, a direct example of information transfer. The increased activity of halogenated derivatives suggests that, as in the cases of algal thyroxine and autonomium in the sponge, discovery of additional compounds with hormonal activity in primitive organisms seems likely. Do predators of these producing organisms enjoy the hormonal advantages of these compounds in analogy to the way in which the sea hare obtains anti-feedants from the algae it devours? Is there a crude analogy wherein the producer of thyroxine or autonomium is as a 'secretory gland' and the predator is as a 'target organ?'

A possible role of metabolites containing heterogeneous or mixed halogens lies in the ability to preferentially remove the most active halogen, for example bromine, leaving behind a chlorinated derivative. If highly specific dehalogenating enzymes have evolved, then friends or enemies might be enzymatically identified by the nature of their heterogeneous halide metabolites or the resultant dehalogenated product. Specific receptor sites may receive the message, with translation to follow. The translation might be an avoidance or attraction reaction.

There is yet another question that may be raised about the possible relationship between halogenation and dehalogenation. A broad range of halogenating and dehalogenating enzymes has been reported in the literature, and much ado is made of the fortuitous and gratuitous nature of these reactions. However, one wonders how many halogenation reactions occur in a biosynthetic sequence followed by a dehalogenation reaction, such as demonstrated with the *in vitro* syntheis of epoxides from alkenes mentioned in Chapter 7. Such a coupled reaction would go unnoticed, since the halogen resides only on a transient compound along a multistep metabolic pathway.

CONCLUSION

The proper recognition of biological halogenation as a legitimate field of natural and applied significance has not yet arrived. It is not around the corner, but it is not too far off. The same respect that halogenation has achieved in organic chemistry will, in time, be forthcoming in biochemistry. Nature has done the experiments. Science has now to understand them; and industry, to profit from them.

Bibliography

Ahern, T. J., G. G. Allan, & D. G. Metcalf (1980) New bromoperoxidases of marine origin. Partial purification and characterization. *Biochim. Biophys. Acta* **616**: 329–339.

Aibara, S., H. Yamashita, E. Mori, M. Kato, & Y. Morita (1982) Isolation and characterization of five neutral isoenzymes of horseradish peroxidase. *J. Biochem.* **92**: 531–539.

Albrich, J. M., C. A. McCarthy, & J. K. Hurst (1981) Biological reactivity of hypochlorous acid: Implications for microbial mechanisms of leukocyte myeloperoxidase. *Proc. Natl. Acad. Sci.* **78**: 210–214.

Albrich, J. M., & J. K. Hurst (1982) Oxidative inactivation of *Escherichia coli* by hypochlorous acid. Rates and differentiation of respiratory from other reaction sites. *FEBS Letters* **144**: 157–161.

Alexander, M. (1977) *Introduction to Soil Microbiology, 2nd Edition*, Chapter 26 Pesticides, pp. 438–456 Wiley, Chichester, England.

Alexander, N. M. (1959) Iodide peroxidase in rat thyroid and salivary glands and its inhibition by antithyroid compounds. *J. Biol. Chem.* **234**: 1530–1533.

Alexander, N. M. (1974) Oxidative cleavage of tryptophanyl peptide bonds during chemical- and peroxidase-catalyzed iodinations. *J. Biol. Chem.* **249**: 1946–1952.

Ali, A., M. Sahai, & A. B. Ray (1984) Physalolactone C, a new withanolide from *Physalis peruviana. J. Natural Prod.* **47**: 648–651.

Allison, N., A. J. Skinner, & R. A. Cooper (1983) The dehalogenases of 2,2-dichloropropionate-degrading bacterium. *J. Gen. Microbiol.* **129**: 1283–1293.

Anantani, Y. S., & B. S. Vaidya (1983) Halotolerance in a blue-green alga: Shifts in electrophoretic patterns of soluble proteins and enzyme peroxidase under salt stress. *Phykos* **22**: 108–112.

Andrews, P. C., & N. I. Krinsky (1981) The reductive cleavage of myeloperoxidase in half, producing enzymically active hemi-myeloperoxidase. *J. Biol. Chem.* **256**: 4211–4218.

Andrews, P. C. & N. I. Krinsky (1982) A kinetic analysis of the interaction of human myeloperoxidase with hydrogen peroxide, chloride ions, and protons. *J. Biol. Chem.* **257**: 13240–13245.

Andrews, P. C., C. Parnes, & N. I. Krinsky (1984) Comparison of myeloperoxidase and hemi-myeloperoxidase with respect to catalysis, regulation, and bactericidal activity. *Arch. Biochem. Biophys.* **228**: 439–442.

Araiso, T. & H. B. Dunford (1981) Effect of modification of heme propionate groups on the reactivity of horseradish peroxidase. *Arch. Biochim. Biophys.* **211**: 346–351.

Araiso, T., R. Rutter, M. M. Palcic, L. P. Hager, & H. B. Dunford (1981) Kinetic analysis of compound I formation and the catalatic activity of chloroperoxidase. *Can. J. Biochem.* **59**: 233–236.

Archer, J. T., G. Air, M. Jackas, & D. B. Morell (1965) Studies on rat eosinophil peroxidase. *Biochem. Biophys. Acta* **99**: 96–101.

Atkin, C. L., M. R. Anderson, & H. J. Eyre (1982) Abnormal neutrophil myeloperoxidase from a patient with chronic myelocytic leukemia. *Arch. Biochem. Biophys.* **214**: 284–292.

Baden, D. G. and M. D. Corbett (1979) Peroxidases produced by the marine sponge *Iotrochota birotulata. Comp. Biochim. Physiol.* **64B**: 279–283.

Bakkenist, A. R. J., R. Weaver, T. Vulsma, H. Plat, & B. F. van Gelder (1978) Isolation procedure and some properties of myeloperoxide from human Leucocytes. *Biochim. Biophys. Acta* **524**: 45–54.

Bakkenist, A. R. J., J. E. G. De Boer, H. Plat, & R. Wever (1980) The halidē complexes of myeloperoxidases and the mechanism of the halogenation reactions. *Biochim. Biophys. Acta* **613**: 337–348.

Bakker, C. N. M. & F. M. Kaspersen (1979) Iodination reactions with lactoperoxidase immobilized on phenoxyacetylcellulose (Enzorb A). *International J. Appl. Radiation Isotopes* **30**: 320.

Bartnicki, E. W. & C. Castro (1969) Biodehalogenation. The pathway for transhalogenation and the stereochemistry of epoxide formation from halohydrins. *Biochem.* **8**: 4677–4680.

Bayse, G. S. & M. Morrison (1971) Peroxidase catalyzed reactions of iodide at low pH. *Arch. Biochem. Biophys.* **145**: 143–148.

Beissner, R. S., W. J. Guilford, R. M. Coates, & L. P. Hager (1981) Synthesis of brominated heptanones and bromoform by a bromoperoxidase of marine origin. *Biochem.* **20**: 3724–3731.

Bennett, L. W., G. W. Miller, M. H. Yu, & R. I. Lynn (1983) Production of fluoroacetate by callus tissue from leaves of *Acacia georginae Fluoride* **16**: 111–117.

Bjorck, K. E. L. (1982) Pharmaceutical preparation with antibacterial effect. Swedish Patent Office, Patent #SE 425,593.

Bjorksten, F. (1968) A kinetic study of the horse-radish peroxidase-catalyzed oxidation of iodide. *Eur. J. Biochem.* **5**: 133–142.

Blackman, A. J. & D. J. Matthews (1982) Halogenated phloroglucinols from *Rhabdonia verticillata. Phytochem.* **21**: 2141–2142.

Blake, C. & B. J. Gould (1984) Use of enzymes in immunoassay techniques. A review. *Analyst* **109**: 533–547.

Boeynaems, J. M. & W. C. Hubbard (1980) Transformation of arachidonic acid into an iodolactone by the rat thyroid. *J. Biol. Chem.* **255**: 9001–9004.

Boeynaems, J. M., D. Reagan, & W. C. Hubbard (1981a) Lactoperoxidase-catalyzed iodination of arachidonic acid: Formation of macrolides. *Lipids* **16**: 246–249.

Boeynaems, J. M., J. T. Watson, J. A. Oates, & W. C. Hubbard (1981b) Iodination of docosahexaenoic acid by lactoperoxidase and thyroid gland *in vitro:* Formation of an iodolactone. *Lipids* **16**: 323–327.

Bolscher, B. G. J. M., H. Plat, & R. Wever (1984a) Some properties of human eosinophil peroxidase, a comparison with other peroxidases. *Biochim. Biophys. Acta* **784**: 177–186.

Bolscher, B. G. J. M., J. R. Zoutberg, R. A. Cuperus, & R. Wever (1984b) Vitamin C stimulates the chlorinating activity of human myeloperoxidase. *Biochim. Biophys. Acta* **784**: 189–191.

Brandange, S., O. Dahlman, A. Mahlen, & L. Morch (1982) Stereoselectivities in enzymatic synthesis of fluorocitric acid. *Acta Chem. Scand.* **B 36**: 67–69.

Brennan, M. R. & K. L. Erickson (1978) Polyhalogenated indoles from the marine alga *Rhodophyllis membrancea* Harvey. *Tetrahed. Letters* 1637–1640.

Breslow, R. (1982) Artificial enzymes. *Science* **218**: 532–537.

Brown, F. S. & L. P. Hager (1967) Chloroperoxidase. IV. Evidence for an ionic electrophilic substitution mechanism. *J. Amer. Chem. Soc.* **89**: 719–720.

Buys, J., R. Wever, R. van Stigt & E. J. Ruitenberg (1981) The killing of newborn larvae of *Trichinella spiralis* by eosinophil peroxidase *in vitro. Eur. J. Immunol.* **11**: 843–845.

Cardellina, J. H., S. B. Horsley, J. Clardy, S. R. Leftow, & J. Meinwald (1982) Secondary metabolites from the red alga *Laurencia intricata:* halogenated enynes. *Can. J. Chem.* **60**: 2675–2677.

Carlson, M. G. C., C. G. B. Peterson, & P. Venge (1985) Human eosinophil peroxidase: Purification and characterization. *J. Immunol.* **134**: 1875–1879.

Carlsson, J. (1983) Catalytic activity of lactoperoxidase in the presence of SCN^-. *Biochem. Biophys. Res. Comm.* **116**: 568–573.

Carlsson, J., Y. Iwami & T. Yamada (1983) Hydrogen peroxide excretion by oral streptococci and effect of lactoperoxidase-thiocyanate-hydrogen peroxide. *Infect. Immun.* **40**: 70–80.

Carlstrom, A. (1969) Physical and compositional investigations of the subfractions of lactoperoxidase. *Acta Chem. Scandinavica* **23**: 185:202.

Carte, B. & D. J. Faulkner (1981) Polybrominated diphenyl ethers from *Dysidea herbacea, Dysidea chlorea and Phyllospongia foliascens. Tetrahed.* **37**: 2335–2339.

Castro, C. E. (1977) Biodehalogenation. *Environ. Health Perspectives* **21**: 279–283.

Castro, C. E. & E. W. Bartnicki (1968) Biodehalogenation. Epoxidation of halohydrins, epoxide opening, and transhalogenation by a *Flavobacterium* sp. *Biochem.* **7**: 3213–3218.

Castro, C. E., R. S. Wade, & N. O. Belser (1985) Biodehalogenation: Reactions of cytochrome P-450 with polyhalomethanes. *Biochem.* **24**: 204–210.

Chacko, C. I., J. L Lockwood, & M. Zabik (1966) Chlorinated hydrocarbon pesticides: Degradation of microbes. *Science* **154**: 893–895.

Chaiet, L., B. H. Arison, R. L. Monaghan, J. P. Springer, J. L. Smith, & S. B. Zimmerman (1984) [R-(Z)-4-amino-3-chloro-pentenedioic acid, a new antibiotic. Fermentation, Isolation and Characterization, *J. Antibiot.* **37**: 207–210.

Chatterjee, D. K., S. T. Kellogg, D. R. Watkins, & A. M. Chakrabarty (1980) Plasmids in the biodegradation of chlorinated aromatic compounds. In *Molecular Biology, Pathogenicity and Ecology of Bacterial Plasmids,* S. B. Levy, R. C. Clowes, and E. L. Koenig (Eds.), Plenum, New York, New York, pp. 519–528.

Chavez, C. & W. H. Flurkey (1984) Bispecific adsorption of peroxidase. *J. Chromatogr.* **298**: 169–171.

Chmielnicka, J., P. I. Ohlsson, K. G. Paul, & T. Stigbrand (1971) Substrate specificity of plant peroxidases. *FEBS Letters* **17**: 181–184.

Clark, A. G. & N. A. Shamaan (1984) Evidence that DDT-dehydrochlorinase from the house fly is a glutathione S-transferase. *Pest. Biochem. Physiol.* **22**: 249–261.

Clark, R. A., S. J. Klebanoff, A. B. Einstein, & A. Fefer (1975) Peroxidase-H_2O_2-halide system: Cytotoxic effect on mammalian tumor cells. *Blood* **45**: 161–170.

Clark, R. A., I. Olsson, & S. J. Klebanoff (1976) Cytotoxicity for tumor cells of cationic proteins from human neutrophil granules. *J. Cell Biol.* **70**: 719–723.

Clark, R. A. & S. J. Klebanoff (1977) Myeloperoxidase-H_2O_2-halide system: Cytotoxic effect on human blood leukocytes. *Blood* **50**: 65–70.

Clark, R. A. & S. J. Klebanoff (1980) Neutrophil-platelet interaction mediated by myeloperoxidase and hydrogen peroxide. *J. Immunol.* **124**: 399–405.

Clark, R. A., S. Szot, K. Venkatasubramanian, & E. Schiffman (1980) Chemotactic factor inactivation by myeloperoxidase-mediated oxidation of methionine. *J. Immunol.* **124**: 2020–2026.

Clark, R. A. (1982) Chemotactic factors trigger their own oxidative inactivation by human neutrophils. *J. Immunol.* **129**: 2725–2728.

Clark, R. A. (1983) Extracellular effects of the myeloperoxidase-hydrogen peroxide-halide system. In *Advances in Inflammation Research, Vol 5,* G. Weissmann, Ed., Raven Press, New York, 107–145.

Clutterbuck, P. W., S. L. Mukhopadhyay, A. E. Oxford, & H. Raistrick (1940) Studies in the biochemistry of micro-organisms. 65 (A) A survey of chlorine metabolism by moulds. (B) Caldariomycin, $C_5H_8O_2Cl_2$, a metabolic product of *Caldariomyces fumago* Woronichin. *Biochem. J.* **34**: 664–677.

Coble, B.-I., C. Dahlgren, J. Hed, & O. Stendahl (1984) Myeloperoxidase reduces the opsonizing activity of immunoglobulin G and complement component C3b. *Biochim. Biophys. Acta* **802**: 501–505.

Colquhoun, E. Q. & R. M. Thomson (1985) Lysosomal thyroid hormone 5^1-deiodinase. *FEBS Letters* **177**: 221–226.

Cooke, J. A., M. S. Johnson, A. W. Davidson, & A. D. Bradshaw (1976) Fluoride in plants colonising fluorospar mine waste in the Peak District and Weardale. *Environ. Pollut.* **11**: 9–23.

Cooney, C. L. & J. Hueter (1974) Enzyme catalysts in the presence of nonaqueous solvents using chloroperoxidase. *Biotechnol. Bioengineer* **16**: 1045–1053.

Corbett, M. D. & B. R. Chipko (1979) Peroxide oxidation of indole to oxindole. by chloroperoxidase catalysis. *Biochem. J.* **183**: 269–276.

Corbett, M. D. B. R. Chipko, & A. O. Batchelor (1980) The action of chloride peroxidase on 4-chloroaniline. *Biochem. J.* **187**: 893–903.

Corbett, M. D. & B. R. Corbett (1983) Enzymatic preparation of [U-^{14}C]-4-chloronitrosobenzene. *Experimentia* **39**: 487–488.

Cramer, R., M. R. Soranzo, P. Dri, R. Menegazzi, A. Pitotti, G. Zabucchi & P. Patriarca (1984a) A simple reliable assay for myeloperoxidase activity in mixed neutrolphil-eosinophil cell suspensions: Application to detection of myeloperoxidase deficiency. *J. Immunological Methods* **70**: 119–125.

Cramer, R., P. Dri, R. Menegazzi & P. Patriarca (1984b) Differential biochemical properties of myeloperoxidase-hydrogen peroxidase-halide system and eosinophil peroxidase-hydrogen peroxidase-halide system. *Acta Med. Rom.* **22**: 162–167.

Cuperus, R. A., A. O. Muijsers, & R. Wever (1983) The effect of D-penicillamine on human myeloperoxidase, a mechanism for the efficacy of the drug in rheumatoid arthritis. *Biochim. Biophys. Acta* **749**: 18–23.

Dalton, H. (1981) Methane monooxygenases from a variety of microbes. In *Microbial Growth on C1 Compounds. Proceedings of the Third International Symposium.* H. Dalton (Ed.), Heyden, London, England, pp. 1–10.

Darkanbaev, T. B., R. M. Kundeva, & B. G. Mukhamedzhanov (1984) Biotechnical aspects in the use of plant peroxidase immobilized on biogel. *Dokl. Vses. Akah. S-kh. Nauk. im V. I. Lenina.* **3**: 20–22.

Davidson B., J. T. Neary, H. V. Strout F. Maloof & M. Soodak (1978) Evidence for a thyroid peroxidase associated "active iodine" species. *Biochim. Biophys. Acta* **522**: 318–326.

Davidson, B., M. Soodak, H. V. Strout, J. T. Neary, C. Nakamura, & F. Maloof (1979) Thiourea and cyanate as inhibitors of thyroid peroxidase: The role of iodide. *Endocrin.* **104**: 919–924.

Davies, D. M., P. Jones, & D. Mantle (1976) The kinetics of formation of horse-radish peroxidase compound I by reaction with peroxobenzoic acids. *Biochem. J.* **154**: 247–253.

De Boer, E., M. C. R. Franssen, H. C. van der Plas, H. Plat, & R. Wever (1985) Bromoperoxidase from marine brown algae: A unique halogenation biocatalyst. *Biocatalysts in Organic Syntheses*, International Symposium of the European Federation of Biotechnology, Netherlands, p. 19.

Deits, T., M. Farrance, E. S. Kay, L. Medhill, E. E. Turner, P. G. Weidman, & B. M. Shapiro (1984) Purification and properties of ovoperoxidase, the enzyme responsible for hardening the fertilization membrane of the sea urchin egg. *J. Biol. Chem.* **259**: 13525–13533.

Diamond, R. D., R. A. Clark, & C. C. Haudenschild (1980) Damage to *Candida albicans* hyphae and psuedohyphae by the myeloperoxidase system and oxidative products of neutrophil metabolism *in vitro*. *J. Clin. Invest.* **66**: 908–917.

Diamond, R. D. & R. A. Clark (1982) Damage to *Aspergillus fumigatus* and *Rhizopus oryzae* hyphae by oxidative and nonoxidative microbicidal products of human neutrophils *in vitro*. *Infect. Immun.* **38**: 487–495.

Dieter, R. K., R. Kinnel, J. Meinwald, & T. Eisner (1979) Brasudol and isobrasudol: Two bromosesquiterpenes from a sea hare (*Aplysia brasiliana*). *Tetrahedr. Letters No. 19:* 1645–1648.

Di Nello, R. K. & D. H. Dophin (1981) Substituted hemins as probes for structure-function relationships in horseradish peroxidase. *J. Biol. Chem.* **256**: 6903–6912.

Dolak, L. A. & C. DeBoer (1980) Clazamycin B is antibiotic 354. *J. Antibiot.* **33**: 83–84.

Drechsel, E. (1896) Contribution to the chemistry of a sea animal. *Z. Biol.* **33**: 85–107.

Dumontet, C. & B. Rousset (1983) Identification, purification, and characterization of a non-heme lactoperoxidase in bovine milk. *J. Biol. Chem.* **258**: 14166–14172.

Dunford, H. B. & J. S. Stillman (1976) On the function and mechanism of action of peroxidases. *Coordination Chem. Rev.* **19**: 187–251.

Dunford, H. B., T. Araiso, D. Job, J. Ricard, R. Rutter, L. P. Hager, R. Wever, W. M. Kast, R. Boelens, N. Ellfolk, & M. Ronnberg (1982) Peroxidases. In *The Biological Chemistry of Iron*, H. B. Dunford *et al.* (Eds.) D. Reidel Publishing Co., 337–355.

Dunford, H. B. & I. M. Ralston (1983) On the mechanism of iodination of tyrosine. *Biochem. Biophys. Res. Comm.* **116**: 639–643.

Edelson, P. J. and Z. A. Cohn (1973) Peroxidase-mediated mammalian cell cytotoxicity. *J. Exp. Med.* **138**: 318–323.

El-Hag, A. & R. A. Clark (1984) Down-regulation of human natural killer activity against tumors by the neutrophil myeloperoxidase system and hydrogen peroxide *J. Immunol.* **133**: 3291–3297.

Elvoson, J. (1974) Biosynthesis of chlorosulfolipids in *Ochromonas danica.* Assembly of the docosane-1,14-diol structure *in vivo*. *Biochem.* **13**: 3483–3487.

Engvild, K. C., H. Egsgaard, & E. Larsen (1981) Determination of 4-chloroindoleacetic acid methyl ester in Vicieae species by gas chromatography-mass spectrometry. *Physiol. Plant.* **53**: 79–81.

Epton, R., M. E. Hobson, & G. Marr (1979) Catalytic activity of poly (acryloyl morpholine) – immobilized horseradish peroxidase in organic/aqueous solvent mixtures. *Enzyme Microb. Technol.* **1**: 37–40.

Erickson, K. L. (1983) Constituents of Laurencia. In *Marine Natural Products: Chem: Biol. Perspect.*, P. J. Scheuer (Ed.) Academic Press, New York, 131–257.

Ezaki, N., T. Shomura, M. Koyama, T. Niwa, M. Kojima, S. Inouye, T. Ito, & T. Niida (1981) New chlorinated nitro-pyrrole antibiotics, pyrrolomycin A and B (SF-2080 A and B), *J. Antibiot.* **34**: 1363–1365.

Ezaki, N., M. Koyama, T. Shomura, T, Tsuruoka, & S. Inouye (1983a) Pyrrolomycins C, D and E, new members of pyrrolomycins. *J. Antibiot.* **36**: 1263–1267.

Ezaki, N., M. Koyama, Y. Kodama, T. Shomura, K. Tashiro, T. Tsuruoaka, S. Inouye, & S-I. Sakai (1983b) Pyrrolomycins F, F_{2a}, F_{2b} and F_3, new metabolites produced by the addition of bromide to the fermentation. *J. Antibiot.* **36**: 1431–1438.

Faulkner, D. J. (1976) Biomimetic synthesis of marine natural products. *Pure & Appl. Chem.* **48**: 25–28.

Faulkner, D. J. (1977) Interesting aspects of marine natural products chemistry. *Tetrahedron* **33**: 1421–1443.

Faulkner, D. J. (1984) Marine natural products: Metabolites of marine algae and herbivorous marine molluscs. *Natural Products Reports* **1**: 251–280.

Fenical, W. (1974) Polyhaloketones from the red seaweed *Asparagopsis taxiformis. Tetrahedr. Letters No. 51/52:* 4463–4466.

Fenical, W. (1975) Halogenation in the Rhodophyta: A Review. *J. Phycol.* **11**: 245–259.

Fenical, W. (1979) Molecular aspects of halogen-based biosynthesis of marine natural products. *Recent Adv. Phytochem.* **13**: 219–239.

Fenical, W, (1982) Natural products chemistry in the marine environment. *Science* **215**: 923–928.

Fiessinger, F., S. W. Maloney, J. Manem, & J. Mallevialle (1984) Potential use of enzymes as catalysts in drinking water for the oxidation of taste causing substances. *Aqua* **2**: 116–118.

Fonong, T. & G. A. Rechnitz (1984) Homogeneous potentiometric enzyme immunoassay for human immunoglobulin G. *Anal. Chem.* **56**: 2586–2590.

Foote, C. S., T. E. Goyne, & R. I. Lehrer (1983) Assessment of chlorination by human neutrophils. *Nature* **301**: 715–716.

Franssen, M. C. R. & H. C. van der Plas (1984) A new enzymatic chlorination of barbituric acid and its 1-methyl and 1,3-dimethyl derivatives. *Recl. Trav. Chim. Pays-Bas* **103**: 99–100.

Gaugler, R. W., P. N. Minke, & J. K. Pecotte (1981) Fluoride accumulation by oral microorganisms. *J. Dent. Res.* **60**: 187–192.

Geigert, J., D. S. Hirano, & M. Moreland (1981) High-performance liquid chromatographic method for the analysis of alkene bromohydrins and iodohydrins. *J. Chromatogr.* **206**: 625–627.

Geigert, J., J. C. Hunter, & S. L. Neidleman (1983a) The discovery of a new ecological niche for halogenating enzymes. In proceedings of the *International Conference on Enzyme Engineering VII*, abstract, page 27.

Geigert, J., S. L. Neidleman, D. J. Dalietos, & S. K. DeWitt (1983b) Haloperoxidases: Enzymatic synthesis of α,β-halohydrins from gaseous alkenes. *Appl. Environ. Microbiol.* **45**: 366–374.

Geigert, J., S. L. Neidleman, D. J. Dalietos, & S. K. DeWitt (1983c) Novel haloperoxidase reaction: Synthesis of dihalogenated products. *Appl. Environ. Microbiol.* **45**: 1575–1581.

Geigert, J., S. L. Neidleman, & D. J. Dalietos (1983d) Novel haloperoxidase substrates. Alkynes and cyclopropanes *J. Biol. Chem.* **258**: 2273–2277.

Geigert, J., D. J. Dalietos, S. L. Neidleman, T. D. Lee, & J. Wadsworth (1983e) Peroxide oxidation of primary alcohols to aldehydes by chloroperoxidase catalysis. *Biochem. Biophys. Res. Comm.* **114**: 1104–1108.

Geigert, J., S. K. DeWitt, S. L. Neidleman, G. Lee, D. J. Dalietos, & M. Moreland (1983f) DMSO is a substrate for chloroperoxidase. *Biochem. Biophys. Res. Comm.* **116**: 82–85.

Geigert, J., S. L. Neidleman, T.-N. Liu, S. K. DeWitt, B. M. Panschar, D. J. Dalietos, & E. R. Siegel (1983g) Production of epoxides from α, β-halohydrins by *Flavobacterium* sp. *Appl. Environ. Microbiol.* **45**: 1148–1149.

Geigert, J., S. L. Neidleman, S. K. DeWitt, & D. J. Dalietos (1984) Halonium ion-induced biosynthesis of chlorinated marine metabolites. *Phytochem. 23*: 287–290.

Geigert, J., T.-N. Liu, & T. M'Timkulu (1985) Stable haloperoxidase and method. U.S. Patent pending.

Ghosal, D., I.-S. Yow, D. K. Chatterjee, & A. M. Chakrabarty (1985) Microbial degradation of halogenated compounds. *Science* **228**: 135–142.

Gjessing, E. C. & J. B. Sumner (1942) Synthetic peroxidases. *Arch. Biochem.* **1**: 1–8.

Gozlan, R. S. & P. Margalith (1974) Iodide oxidation by *Pseudomonas iodooxidans. J. Appl. Bact.* **37**: 493–499.

Gregson, R. P., B. A. Baldo, P. G. Thomas, R. J. Quinn, P. R. Bergquist, J. F. Stephens, & A. R. Horne (1979) Fluorine is a major constituent of the marine sponge *Halichondria moorei. Science* **206**: 1108–1109.

Griffin, B. W. & P. L. Ashley (1984) Evidence for a radical mechanism of halogenation of monochlorodimedone catalyzed by chloroperoxidase. *Arch. Biochem. Biophys.* **233**: 188–196.

Grime, J. K. & K. R. Lockhart (1979) The determination of iodide in the presence of other halides by enzymatic enthalpimetry. *Analytica Chim. Acta* **108**: 363–366.

Grisham, M. B., M. M. Jefferson, D. F. Metton, & E. L. Thomas (1984) Chlorination of endogenous amines by isolated neutrophils. Ammonia-dependent bactericidal, cytotoxic and cytolytic activities of the chloramines. *J. Biol. Chem.* **259**: 10404–10413.

Gschwend, P. M., J. K. MacFarlane, & K. A. Newman (1985) Volatile halogenated organic compounds released to seawater from temperate marine microalgae. *Science* **227**: 1033–1035.

Gupta, R. K., A. S. Mildvan, & G. R. Schonbaum (1980) Water proton relaxation studies of the heme-environment in Mn(III)-substituted and native horseradish peroxidases. *Arch. Biochem. Biophys.* **202**: 1–7.

Gusakova, S. D. & A. U. Umarov (1977) Dibromo- and tetrabromostearic acids of the oil from *Eremostachys molucelloides* seeds. *Chem. Abstracts* **86**: 103053z.

Haas, D. (1983) Genetic aspects of biodegradation by pseudomonads. *Experientia* **39**: 1199–1213.

Hager, L. P., D. R. Morris, F. S. Brown, & H. Eberwein (1966) Chloroperoxidase. II. Utilization of halogen anions. *J. Biol. Chem.* **241**: 1769–1776.

Hager, L. P., J. A. Thomas, & D. R. Morris (1970) Studies on the relationship of chloroperoxidase-halide and chloroperoxidase-hydrogen peroxide complexes to the mechanism of the halogenation reaction. In *Biochemistry of the Phagocytic Process*, J. Schultz, (Ed.), North-Holland Publishing Co., North Holland, 67–87.

Hager, L. P. (1982) Mother nature likes some halogenated compounds. *Basic Life Sci.* **19**: 415–429.

Haines, T. H. (1973) Halogen- and sulfur-containing lipids of *Ochromonas. Ann. Rev. Microbiol.* **27**: 403–411.

Hall, R. J. (1977) The presence and biosynthesis of carbon-fluorine compounds in tropical plants and soils. In Trace *Substances in Environmental Health XI*, D. D. Hemphill, (Ed.), Univ. Missouri, 156–163.

Hankin, L. & B. L. Sawhney (1984) Microbial degradation of polychlorinated biphenyls in soil. *Soil Science* **137**: 401–407.

Harrison, J. E. & J. Schultz (1976) Studies on the chlorinating activity of myeloperoxidase. *J. Biol. Chem.* **251**: 1371–1374.

Harrison, J. E. (1982) The role of peroxide in the functional mechanism of myeloperoxidase. In *Oxidases and Related Redox Systems, Proceedings of 3rd International Symposium*, T. King, H. S. Mason and M. Morrison (Eds.), Pergamon, New York, New York, 717–732.

Hasegawa, E. & T. Kobayashi (1983) Method for recovering myeloperoxidase and pharmaceutical composition containing myeloperoxidase as major constituent. U.S. Patent Office, Patent #4,379,141.

Hashimoto, A. & M. A. Pickard (1984) Chloroperoxidases from *Caldarimyces* (=*Leptoxyphium*) cultures: glycoproteins with variable carbohydrate content and isoenzymic forms. *J. Gen. Microbiol.* **130**: 2051–2058.

Henderson, W. R., A. Joerg, & S. J. Klebanoff (1981) Slow-reacting substance generation by eosinophils and its inactivation by peroxidase. *Clin. Res.* **29**: 491A.

Hewson, W. D. & H. B. Dunford (1976) Stoichiometry of the reaction between horseradish peroxidase and p-cresol. *J. Biol. Chem.* **251**: 6043–6052.

Hewson, W. D. & L. P. Hager (1979) Mechanism of the chlorination reaction catalyzed by horseradish peroxidase with chlorite. *J. Biol. Chem.* **254**: 3175–3181.

Hewson, W. D. & L. P. Hager (1980) Bromoperoxidases and halogenated lipids in marine algae. *J. Phycol.* **16**: 340–345.

Hirano, D. S., J. Geigert, & T. D. Lee (1983) Fused silica capillary GC/MS analysis of heterogeneous dihalogenated 1-propanols. *J. High Resolut. Chromatogr. and Chromatogr. Communicat.* **6**: 38–39.

Holzbach, R., H. Pape, D. Hook, E. F. Kreutzer, C. Chang, & H. G. Floss (1978) Biosynthesis of the macrolide antibiotic chlorothricin: Basic building blocks. *Biochem.* **17**: 556–560.

Honka, E., P. I. Ohlsson, & K. G. Paul (1982) The adsorption of lactoperoxidase to glass. *Acta Chem. Scandinavica B* **36**: 273–274.

Horowitz, A., J. M. Suflita, & J. M. Tiedje (1983) Reductive dehalogenations of halobenzoates by anaerobic lake sediment microorganisms. *Appl. Environ. Microbiol.* **45**: 1459–1465.

Hoyle, M. C. (1977) High resolution of peroxidase-indoleacetic acid oxidase isoenzymes from horseradish by isoelectric focusing. *Plant Physiol.* **60**: 787–793.

Hunt, S. (1984) Halogenated tyrosine derivatives in invertebrate scleroproteins: Isolation and identification. Ch. 27. In *Methods of Enzymology Volume 107 Posttranslational Modifications Part B*, F. Wold and K. Moldave (Ed.), Academic Press, New York, 413–438.

Hunter, J. C., M. Fonda, L. Sotos, B. Toso, & A. Belt (1985a) Ecological approaches to isolation. *Developments Industrial Microbiol.*, in press.

Hunter, J. C., A. Belt, L. Sotos, & M. Fonda (1985b) Fungal chloroperoxidase method. U.S. Patent pending.

Hurst, J. K., J. M. Albrich, T. R. Green, H. Rosen, & S. Klebanoff (1984) Myeloperoxidase – dependent fluorescein chlorination by stimulated neutrophils. *J. Biol. Chem.* **259**: 4812–4821.

Huwiler, M. & H. Kohler (1984) Pseudo-catalytic degradation of hydrogen peroxide in the lactoperoxidase /H_2O_2/iodide system. *Eur. J. Biochem.* **141**: 69–74.

Hylin, J. W., R. E. Spenger, & F. A. Gunther (1969) Potential interferences in certain pesticide residue analyses from organochlorine compounds occurring naturally in plants. *Residue Rev.* **26**: 127–138.

Iizuka, T., K. Watanabe, R. Makino, K. Sakaguchi, F. Mitani, Y. Ishimura, K. Kawabe, Z. Yoshida, & H. Ogoshi (1982) Effects of heme substitution on the co-binding of myoglobin and horseradish peroxidase. In *Oxygenases and Oxygen Metabolism*, M. Nozaki, S. Yamamoto, Y. Ishimura, M. J. Coon, L. Ernster, R. W. Estabrook, (Eds.), Academic Press, New York, 445–450.

Ireland, C., M. O. Stallard, & D. J. Faulkner (1976) Some chemical constituents of the digestive gland of the sea hare *Aplysia californica*. *J. Org. Chem.* **41**: 2461–2465.

Izac, R. R. & J. J. Sims (1979) Marine nature products. 18. Iodinated sesquiterpenes from the red algal genus *Laurencia*. *J. Amer. Chem. Soc.* **101**: 6136–6137.

Janssen, D. B., A. Scheper, L. Dijkhuizen, & B. Witholt (1985) Degradation of halogenated aliphatic compounds by *Xanthobacter autotrophicus* GJ10. *Appl. Environ. Microbiol.* **49**: 673–677.

Jasin, H. E. (1983) Generation of IgG aggregates by the myeloperoxidase-hydrogen peroxide system. *J. Immunol.* **130**: 1918–1923.

Jenkins, I. D., J. P. H. Verheyden, and J. G. Moffatt (1976) 4′-Substituted nucleosides. 2. Synthesis of the nucleoside antibiotic nucelocidin. *J. Amer. Chem. Soc.* **98**: 3346–3357.

Jerina, D., G. Guroff, & J. Daly (1968) Enzymic and nonenzymic hydroxylation and chlorination of p-deuteroanisole. *Arch. Biochem. Biophys.* **124**: 612–615.

Johnson, M. A. & R. Croteau (1984) Biosynthesis of ascaridole: Iodide peroxidase – catalyzed synthesis of a monoterpene endoperoxide in soluble extracts of *Chenpodium ambrosiordes* fruit. *Arch. Biochem. Biophys.* **235**: 254–266.

Johnston, H. W., G. G. Briggs, & M. Alexander (1972) Metabolism of 3-chlorobenzoic acid by a pseudomonad. *Soil Biol. Biochem.* **4**: 187–190.

Johnston, J. B. & S. G. Robinson (1984) Pollution technology review. No. 106: Genetic engineering and new pollution control technologies. Noyes Data Corp., Park Ridge, New Jersey, 131 pp.

Jones, B. A., I. J. Tinsley and R. R. Lowry (1983a) Bromine levels in tissue lipids of rats fed brominated fatty acids. *Lipids* **18**: 319–316.

Jones, B. A., I. J. Tinsley, G. Wilson, & R. R. Lowry (1983b) Toxicology of brominated fatty acids: Metabolite concentration and heart and liver changes. *Lipids* **18**: 327–334.

Jong, E. C., W. R. Henderson, & S. J. Klebanoff (1980) Bactericidal activity of eosinophil peroxidase. *J. Immunol.* **124**: 1378–1382.

Jong, E. C. & S. J. Klebanoff (1980) Eosinophil-mediated mammalian tumor cell cytotoxicity: Role of the peroxidase system. *J. Immunol.* **124**: 1949–1953.

Jong, E. C., A. A. F. Mahmoud, & S. J. Klebanoff (1981) Peroxidase-mediated toxicity to schistosomula of *Schistosoma mansoni*. *J. Immunol.* **126**: 468–471.

Kanner, J. & J. E. Kinsella (1983) Lipid deterioration initiated by phagocytic cells in muscle foods: β-carotene destruction by a myeloperoxidase-hydrogen peroxide-halide system. *J. Agric. Food Chem.* **31**: 370–376.

Kanofsky, J. R. (1984) Singlet oxygen production by chloroperoxidase-hydrogen peroxide-halide systems. *J. Biol. Chem.* **259**: 5596–5600.

Kanofsky, J. R., J. Wright, G. E. Miles-Richardson, & A. I. Tauber (1984) Biochemical requirements for singlet oxygen production by purified human myeloperoxidase, *J. Clin. Invest.* **74**: 1489–1495.

Karonen, S.-L. (1981) Solid lactoperoxidase in the iodination of L-tyrosine and albumin. *Appl. Biochem. Biotechnol.* **6**: 119–128.

Karns, J. S., J. J. Kilbane, D. K. Chatterjee, & M. M. Chakrabarty (1984) Microbial biodegradation of 2,4,5-trichlorophenoxy acetic acid and chlorophenols *Basic Life Sci.* **28**: 3–21.

Katayama, A., K. Ina, H. Nozaki, & M. Nakayama (1982) Structural elucidation of kurodainol, a novel halogenated monoterpene from sea hare (*Aplysia kurodai*). *Agric. Biol. Chem.* **46**: 859–860.

Kaul, P. N. (1982) Biomedical potential of the sea. *Pure Appl. Chem.* **54**: 1963–1972.

Kawasaki, H., K. Miyoshi & K. Tonomura (1981a) Purification, crystallization and properties of haloacetate halidohydrolase from *Pseudomonas* species. *Agric. Biol. Chem.* **45**: 543–544.

Kawasaki, H., N. Tone, & K. Tonomura (1981b) Purification and properties of haloacetate halidohydrolase specified by plasmid from *Moraxella* sp. strain B. *Agric. Biol. Chem.* **45**: 35–42.

Kazlauskas, R., P. T. Murphy, R. J. Wells, J. A. Baird-Lambert, & D. D. Jamieson (1983) Halogenated pyrrolo[2,3-d]pyrimidine nucleosides from marine organisms. *Aust. J. Chem.* **36**: 165–170.

Kedderis, G. L. & P. F. Hollenberg (1984) Peroxidase-catalyzed N-demethylation reactions. Substrate deuterium isotope effects. *J. Biol. Chem.* **259**: 3663–3668.

Kersten, H. W., W. R. Moorer, & R. Wever (1981) Thiocyanate as a cofactor in myeloperoxidase activity against *Streptococcus mutans. J. Dent. Res.* **60**: 831–837.

Kessler, J. H. & R. S. Rosenbaum (1984) Bactericidal method. U.S. Patent Office, Patent #4,476,108.

Kinnel, R., A. J. Duggan, T. Eisner, & J. Meinwald (1977) Panacene: An aromatic bromoallene from a sea hare (*Aplysia brailiana*). *Tetrahedr. Letters* **44**: 3913–3916.

Kinnel, R. B., R. K. Dieter, J. Meinwald, D. Van Engen, J. Clardy, T. Eisner, M. O. Stallard, & W. Fenical (1979) Brasilenyne and *cis*-dihydrorhodophytin: Antifeedant medium-ring haloethers from a sea hare (*Aplysia brasiliana*). *Proc. Natl. Acad. Sci. USA* **76**: 3576–3579.

Klages, V. & F. Lingens (1979) Degradation of 4-chlorobenzoic acid by a *Nocardia* species. *FEMS Microbiol. Letters* **6**: 201–203.

Klebanoff, S. J. (1967) Iodination of bacteria: A bactericidal mechanism. *J. Exp. Med.* **126**: 1063–1079.

Klebanoff, S. J. & R. A. Clark (1975) Hemolysis and iodination of erythrocyte components by a myeloperoxidase-mediated system. *Blood* **45**: **699–707**.

Klibanov, A. M. & E. D. Morris (1981) Horseradish peroxidase for the removal of carcinogenic aromatic amines from water. *Enzyme Microb. Technol.* **3**: 119–122.

Klibanov, A. M., Z. Berman, & B. N. Alberti (1981) Preparative hydroxylation of aromatic compounds catalyzed by peroxidase. *J. Amer. Chem. Soc.* **103**: 6264–6265.

Klibanov, A. M., T.-M. Tu, & K. P. Scott (1983) Peroxidase-catalyzed removal of phenols from coal-conversion waste waters. *Science* **221**: 259–261.

Klyachko, N. L., A. V. Levashov, & K. Martinek (1984). Catalysis by enzymes entrapped into reversed micelles of surfactants in organic solvents. Peroxidase in the aerosol OT-water-octane system. *Mol. Biol. (Moscow)* **18**: 1019–1031.

Kobayashi, M., T. Tanaka, & T. Usui (1982) Inactivation of lysosomal enzymes by the respiratory burst of polymorphonuclear leukocytes. Possible involvement of myeloperoxidase -H_2O_2-halide system. *J Lab. Clin. Med.* **100**: 986–907.

Kobayashi, H., H. Karasawa, T. Miyase, & S. Fukushima (1984) Studies on the constituents of *Cistanchis* herba. II. Isolation and structures of new iridoids, cistanin and cistachlorin. *Chem. Pharm. Bull.* **32**: 1729–1734.

Kohler-Staub, D. & T. Leisinger (1985) Dichloromethane dehalogenase of *Hyphomicrobium* sp. strain DM2. *J. Bacteriol.* **162**: 676–681.

Kollonitsch, J., S. Marburg, & L. M. Perkins (1970) Enzymatic formation of chiral structures in racemic form. *J. Amer. Chem. Soc.* **92**: 4489–4490.

Koyama, M., Y. Kodama, T. Tsuruoka, N. Ezaki, T. Niwa, & S. Inouye (1981) Structure and synthesis of pyrrolomycin A, a chlorinated nitro-pyrrole antibiotic. *J. Antibiotics* **34**: 1569–1576.

Kozarich, J. W., R. V. J. Chari, J. C. Wu, & T. L. Lawrence (1981) Fluoromethylglyoxal: Synthesis and glyoxalase I catalyzed product partitioning via a presumed enediol intermediate. *J. Amer. Chem. Soc.* **103**: 4593–4595.

Kwasniewska, K. & K. L. E. Kaiser (1983) Toxicities of selected phenols to fermentative and oxidative yeasts. *Bull. Environ. Contam. Toxicol.* **31**: 181–194.

Laane, C., W. Pronk, M. Franssen & C. Veeger (1984) Use of a bioelectrochemical cell for the synthesis of (bio) chemicals. *Enzyme Microb. Technol.* **6**: 165–168.

Laane, C., A. Weijland, M. Franssen & C. Veeger (1985) Bioelectrochemical production of fine chemicals using chloroperoxidase. *Biocatalysts in Organic Syntheses*, International Symposium of the European Federation of Biotechnology, Netherlands, 31.

Lal, R. & D. M. Saxena (1982) Accumulation, metabolism, and effects of organochlorine insecticides on microorganisms. *Microbiol. Rev.* **46**: 95–127.

Lambeir, A.-M. & H. B. Dunford (1983) A steady state kinetic analysis of the reaction of chloroperoxidase with peracetic acid, chloride, and 2-chlorodimedone. *J. Biol. Chem.* **258**: 13558–13563.

Lane, T. A. & G. E. Lamkin (1983) Myeloperoxidase-mediated modulation of chemotactic peptide binding to human neutrophils. *Blood* **61**: 1203–1207.

Lee, T. D., J. Geigert, D. J. Dalietos, & D. S. Hirano (1983a) Neighboring group migration in enzyme-mediated halohydrin formation. *Biochem. Biophys. Res. Comm.* **110**: 880–883.

Lee, C. W., R. A. Lewis, A. I. Tauber, M. Mehrotra, E. J. Corey, & K. F. Austen (1983b) The myeloperoxidase-dependent metabolism of leukotrienes C_4, D_4, and E_4 to 6-*trans*-leukotriene B_4 diastereoisomers and the subclass-specific *S*-diastereoisomeric sulfoxides. *J. Biol. Chem.* **258**: 15004–15010.

Lehrbach, P. R., J. Zeyer, W. Reineke, H.-J. Knackmuss, & K. N. Timmus (1984) Enzyme requirement in vitro: Use of cloned genes to extend the range of haloaromatics degraded by *Pseudomonas* sp. strain B13. *J. Bacteriol.* **158**: 1025–1032.

Leisinger, T. (1983) Microorganisms and xenobiotic compounds. *Experientia* **39**: 1183–1191.

Levine, S. D., S. L. Neidleman, & M. Oberc (1968) An enzymatic route to α-bromo steriodal ketones. *Tetrahedr.* **24**: 2979–2984.

Le-Van, N. & S. J. Wratten (1984) Compound 30.4, an unusual chlorinated 1,4-benzoxazin-3-one derivative from corn (*Zea mays*). *Tetrahedr. Letters* **25**: 145–148.

Libby, R. D., J. A. Thomas, L. W. Kaiser, & L. P. Hager (1982) Chloroperoxidase halogenating reactions. Chemical versus enzymatic halogenating intermediates. *J. Biol. Chem.* **257**: 5030–5037.

Lin, L.-J. and G. A. Cordell (1984) Synthesis of coumarinolignans through chemical and enzymic oxidation. *J. Chem. Soc. Chem. Commun.*, 160–161.

Loo, T. L., J. W. Burger, & R. H. Adamson (1964) Bromination of phthalein dyes by the uterus of the dogfish, *Squalus acanthias. Proc. Soc. Exp. Biol. Med.* **114**: 60–63.

Luiken, J., R. Van der Zee, & G. W. Welling (1984) Structure and activity of proteins after reversed-phase high-performance liquid chromatography. *J. Chromatogr.* **284**: 482–486.

Magnusson, R. P., A. Taurog, & M. L. Davis (1984) Mechanisms of thyroid peroxidase- and lactoperoxidase-catalyzed reactions involving iodide. *J. Biol. Chem.* **259**: 13783–13790.

Manthey, J. A. & L. P. Hager (1981) Purification and properties of bromoperoxidase from *Penicillus capitatus. J. Biol. Chem.* **256**: 11232–11238.

Manthey, J. A., L. P. Hager, & K. D. McElvany (1984) Protein bromination by bromoperoxidase from *Penicillus capitatus*, Ch. 28. In *Methods of Enzymology Volume 107 Posttranslational Modifications Part B*, F. Wold and K. Moldave (Ed.), Academic Press, New York, 439–445.

Marinchenko, G. V. (1981) Isolation of ^{125}I-prolactin using lactoperoxidase sorbed on zirconium phosphate. *Probl. Endokrinol.* **27**: 54–58.

Marks, T. S., R. Wait, A. R. W. Smith, & A. V. Quirk (1984) The origin of the oxygen incorporated during the dehalogenation/hydroxylation of 4-chlorobenzoate by an *Arthrobacter* sp. *Biochem. Biophys. Res. Comm.* **124**: 669–674.

Markus, A., V. Klages, S. Krauss, & F. Lingens (1984) Oxidation and dehalogenation of 4-chlorophenylacetate by a two-component enzyme system from *Pseudomonas* sp. strain CBS3, *J. Bacteriol.* **160**: 618–621.

Marletta, M. A., Y.-F. Cheung, & C. Walsh (1982) Stereochemical studies on the hydration of monofluorofumarate and 2,3-difluorofumarate by fumarase. *Biochem.* **21**: 2637–2644.

Marais, J. S. C. (1944) Monofluoroacetic acid, the toxic principle of "Gifblaar" *Dichapetalum cymosum* (Hook) Engl. *Onderstepoort J. Vet Sci. Animal Industry* **20**: 67–73.

Martinek, K., A. V. Levashov, Y. L. Khmelnitsky, N. L. Klyachko, & I. V. Berezin (1982) Colloidal solution of water in organic solvents: A microheterogeneous medium for enzymatic reactions. *Science* **218**: 889–891.

Marshall, V. M. E. & B. Reiter (1980) Comparison of the antibacterial activity of the hypothiocyanite anion towards *Streptococcus lactis* and *Escherichia coli. J. Gen. Microbiol.* **120**: 513–516.

Martyn, R. P., S. C. Branzer, & G. T. Sperl (1981) Antimicrobial chlorinated β-diketones. *Bios* **52**: 8–12.

Mason, C. P., K. R. Edwards, R. E. Carlson, J. Pignatello, F. K. Gleason, & J. M. Wood (1982) Isolation of chlorine-containing antibiotic from the freshwater cyanobacterium *Scytonema hofmanni. Science* **215**: 400–402.

Matheson, N. R., P. S. Wong, & J. Travis (1981) Isolation and properties of human neutrophil myeloperoxidase. *Biochem.* **20**: 325–330.

Matheson, N. R. (1982) The effect of antiarthritic drugs and related compounds on the human neutrophil myeloperoxidase system. *Biochem. Biophys. Res. Comm.* **108**: 259–265.

Matkovics, B., Z. Rakonczay, S. E. Rajki, & L. Balaspiri (1971) Steroids. XII. Iodination of aromatic steroids by peroxidases. *Steroidologia* **2**: 77–79.

McConnell, O. J. & W. Fenical (1980) Halogen chemistry of the red alga *Bonnemaisonia. Phytochem.* **19**: 233–247.

Mead, R. J. & W. Segal (1972) Fluoroacetic acid biosynthesis: A proposed mechanism. *Aust. J. Biol. Sci.* **25**: 327–333.

Mercer, E. I. & C. L. Davies (1979) Distribution of chlorosulpholipids in algae. *Phytochem.* **18**: 457–462.

Merrill, D. P. (1980) Purification of human myeloperoxidase by concanavalin A-Sepharose affinity chromatography. *Preparative Biochem.* **10**: 133–150.

Mickelson, M. N. (1979) Antibacterial action of lactoperoxidase-thiocyanate-hydrogen peroxide on *Streptococcus agalactiae. Appl. Environ. Microbiol.* **38**: 821–826.

Miller, J. N., B. F. Rocks, & D. T. Burns (1976) Preparation and properties of horseradish peroxidase bound covalently to polystyrene beads. Application in the semi-automatic determination of hydrogen peroxide with homovanillic acid as substrate. *Analyt. Chim. Acta* **86**: 93–101.

Minale, L. & R. Riccio (1976) Constituents of the digestive gland of the molluscs of the genus *Aplysia* – I. Novel diterpenes from *Aplysia depilans. Tetrahedr. Letters No.* **31**: 2711–2714.

Mitchell, E. D., Jr., R. Loring, K. L. Carraway, & B. Cunningham (1981) Oxidative inactivation of *B. subtilis* α-amylase by chloroperoxidase. *Proc. Okla. Acad. Sci.* **61**: 43–48.

Moore, R. E. (1978) Halogenated compounds from the *Bonnemaisoniaceae.* In *Marine Natural Products, Chemical and Biological Perspectives*, P. J. Scheuer, (Ed.), Academic Press, New York, 59–73.

Moore, R. E. (1979) Marine aliphatic natural products. *Aliphatic and Related Natural Product Chemistry*, **1**: 20–67.

Moore, R. E., C. Cheuk, & G. M. L. Patterson (1984) Hapalindoles: new alkaloids from the blue-green alga *Hapalosiphon fontinalis*. *J. Amer. Chem. Soc.* **106**: 6456–6457.

Morishima, I, Y. Shiro, & Y. Takamuki (1983) NMR studies of ruthenium(II) porphyrin π-cation radicals with relevance to horseradish peroxidase compound I. *J. Am. Chem. Soc.* **105**: 6168–6170.

Morris, D. R. & L. P. Hager (1966a) Mechanism of the inhibition of enzymatic halogenation by anti-thyroid agents. *J. Biol. Chem.* **241**: 3582–3589.

Morris, D. R. & L. P. Hager (1966b) Chloroperoxidase I. Isolation and properties of the crystalline glycoprotein. *J. Biol. Chem.* **241**: 1763–1768.

Morrison, M. & G. R. Schonbaum (1976) Peroxidase-catalyzed halogenation. *Ann. Rev. Biochem.* **45**: 861–888.

Motosugi, K., N. Esaki, & K. Soda (1982a) Purification and properties of a new enzyme, DL-2-haloacid dehalogenase, from *Pseudomonas* sp. *J. Bacteriol.* **150**: 522–527.

Motosugi, K., N. Esaki, & K. Soda (1982b) Purification and properties of 2-haloacid dehalogenase from *Pseudomonas putida*. *Agric. Biol. Chem.* **46**: 837–838.

Motosugi, K. & K. Soda (1983) Microbial degradation of synthetic organochlorine compounds. *Experientia* **39**: 1214–1220.

Murty, K. V. S. S., K. E. Rao, & C. S. Sastry (1984) New analytical assays using hydrogen peroxide and peroxidase. *Analyst* **109**: 405–406.

Mynderse, J. S. & R. E. Moore (1978) Toxins from blue-green algae: structures of oscillatoxin A and three related bromine-containing toxins. *J. Org. Chem.* **43**: 2301–2303.

Nakagawa, H., T. Kotani, S. Ohtaki, M. Nakamura, & I. Yamazaki (1985) Purification of thyroid peroxidase by monoclonal antibody-assisted immunoaffinity chromatography. *Biochem. Biophys. Res. Comm.* **127**: 8–14.

Nakamura, M., I. Yamazaki, H. Nakagawa, & S. Ohtaki (1983) Steady state kinetics and regulation of thyroid peroxidase-catalyzed iodination. *J. Biol. Chem.* **258**: 3837–3842.

Narayanan, S., M. R. S. Iyengar, P. L. Ganju, S. Rengarju, T. Shomura, T. Tsuruoka, S. Inouye, & T. Niida (1980) γ-Chloronovaline, a leucine analog from *Streptomyces*. *J. Antibiot.* **33**: 1249–1255.

Naylor, S., F. J. Hanke, L. V. Manes, & P. Crews (1983) Chemical and biological aspects of marine monoterpenes. *Prog. Chem. Org. Nat. Prod.* **44**: 189–241.

Neary, J. T., M. Soodak, & F. Maloof (1984) Iodination by thyroid peroxidase. Ch. 29. In *Methods of Enzymology Volume 107 Post translational Modifications Part B*, F. Wold and K. Moldave (Ed.), Academic Press, New York, 445–475.

Neidleman, S. L., P. A. Diassi, B. Junta, R. M. Palmere, & S. C. Pan (1966) The enzymatic halogenation of steroids. *Tetrahedr. Letters* **44**: 5337–5342.

Neidleman, S. L. & M. A. Oberc (1968) Enzymatic bromination of 16-dehydroestradiol 3-methyl ether 17-acetate to 16α-bromoestrone 3-methyl ether. *J. Bacteriol.* **95**: 2424–2425.

Neidleman, S. L. & S. D. Levine (1968) Enzymatic bromohydrin formation. *Tetrahedr. Letters*, 4057–4059.

Neidleman, S. L., A. I. Cohen, & L. Dean (1969) Enzymatic bromination of the thiazole ring. *Biotech. Bioengineer.* **II**: 1227–1232.

Neidleman, S. L. (1974) Process for the production of antimicrobial agents. U.S. Patent Office, Patent #3,817,835.

Neidleman, S. L. (1975) Microbial halogenation. *CRC Crit. Rev. Microbiol.* **5**: 333–358.

Neidleman, S. L. (1980) Use of enzymes as catalysts for alkene oxide production. *Hydrocarbon Processing*, 135–138.

Neidleman, S. L., J. Geigert, W. F. Amon, E. Liu, & B. Wolf (1980) A unique multi-enzyme process for the production of commodity chemicals. *Biochemical Apporaches to Organic Synthesis*, International Symposium of the European Biochemical Societies, FEBS, Israel.

Neidleman, S. L. & J. Geigert (1981) Method for producing iodine. U.S. Patent Office, Patent #4,282,324.

Neidleman, S. L., W. F. Amon Jr. & J. Geigert (1981) Preparation of epoxides and glycols from gaseous alkenes. U.S. Patent Office, Patent #4,284,723.

Neidleman, S. L. & J. Geigert (1983a) The enzymatic synthesis of heterogeneous dihalide derivatives: A unique biocatalytic discovery. *Trends in Biotechnol.* **1**: 1–5.

Neidleman, S. L. & J. Geigert (1983b) Biological halogenation and epoxidation. *Biochem. Soc, Symp.* **48**: 39–52.

Neidleman, S. L. & J. Geigert (1984) Reaction conditions: Determinants in biocatalytic expression. *Dev. Indust. Microbiol.* **25**: 19–30.

Nettleton, D. E., W. T. Bradner, J. A. Bush, & T. W. Doyle (1984) European Patent Office, Patent #0115350A2.

Nicholls, P. (1962) Peroxidase as an oxygenase. In *Oxygenases*, O. Hayaishi (Ed.), Academic Press, New York, 273–305.

Nittala, S. S., V. V. Velde, F. Frolow, & D. Lavie (1981) Chlorinated withanolides from *Withania somnifera* and *Acnistus breviflorus*. *Phytochem.* **20**: 2547–2552.

Norton, R. S. & R. J. Wells (1982) A series of chiral polybrominated biindoles from the marine blue-green alga *Rivularia firma*. Application of ^{13}C NMR spin-lattice relaxation data and ^{13}C-^{1}H coupling constants to structure elucidation. *J. Amer. Chem. Soc.* **104**: 3628–3635.

Nunez, J. (1984) Thyroid hormones: Mechanism of phenoxy ether formation, Ch. 30. In *Methods of Enzymology Volume 107 Posttranslational Modifications Part B*, F. Wold and K. Moldave (Ed.), Academic Press, New York, 476–488.

Odajima, T. (1981) Oxidative destruction of the microbial metabolite aflatoxin by the myeloperoxidase-hydrogen peroxide-chloride system. *Archs. Oral Biol.* **26**: 339–340.

Ogata, Y., M. Kimura, Y. Kondo, H. Katoh, & F.-C. Chen (1984) Orientation in the chlorination of phenol and of anisole with sodium and t-butyl hypochlorites in various solvents. *J. Chem. Soc. Perkin Trans.* **II**, 451–453.

Ogawa, S., Y. Shiro, & I. Morishima (1979) Calcium binding by horseradish peroxidase c and the heme environmental structure. *Biochem. Biophys. Res. Comm.* **90**: 674–678.

Ohlsson, P.-I. (1984) Lactoperoxidase, a dithionite ion dismutase. *Eur. J. Biochem.* **142**: 233–238.

Ohtaki, S., H. Nakagawa, S. Nakamura, M. Nakamura, & I. Yamazaki (1985) Characterization of hog thyroid peroxidase. *J. Biol. Chem.* **260**: 441–448.

Olsen, R. L. & C. Little (1983) Purification and some properties of myeloperoxidase and eosinophil peroxidase from human blood. *Biochem. J.* **209**: 781–787.

Olsen, R. L. & C. Little (1984) Studies on the subunits of human myeloperoxidase. *J. Biochem.* **222**: 701–709.

Olsson, B. & L. Ogren (1983) Optimization of peroxidase immobilization and of the design of packed-bed enzyme reactions for flow injection analysis. *Analyt. Chim. Acta.* **145**: 87–99.

Orndorff, S. A. (1983) Enzymatic catalyzed biocide system. U.S. Patent Office, Patent #4,370,199.

Paice, M. G. & L. Jurasek (1984) Peroxidase-catalyzed color removal from bleach plant effluent. *Biotechnol. Bioengineer.* **26**: 477–480.

Patterson, E. L., W. W. Andres, & L. A. Mitscher (1967) Isolation of the bromo analogue of caldariomycin from *Caldariomyces fumago. Appl. Microbiol.* **15**: 528–530.

Paul, B. B., A. A. Jacobs, R. R. Strauss, & A. J. Sbarra (1970) Role of the phagocyte in host-parasite interactions. XXIV. Aldehyde generation by the myeloperoxidase-H_2O_2-chloride antimicrobial system: A possible *in vivo* mechanism of action. *Infect. Immun.* **2**: 414–418.

Paul, K. G., P. I. Ohlsson, & S. Wold (1979) Formation of horseradish peroxidase compound I with alkyl hydroperoxides. *Acta Chem. Scand. B* **33**: 747–754.

Paul, V. J., H. H. Sun, & W. Fenical (1982) Udoteal, a linear diterpenoid feeding deterrent from the tropical green alga *Udotea flabellum. Phytochem.* **21**: 468–469.

Pedersen, M. (1976) A brominating and hydroxylating peroxidase from the red alga *Cystoclonium purpureum. Physiol. Plant.* **37**: 6–11.

Pedersen, M. & E. J. Da Silva (1973) Simple brominated phenols in the bluegreen alga *Calothrix brevissima* West. *Plants* **115**: 83–86.

Pekoe, G., K. Van Dyke, D. Peden, Mengoli, & D. English (1982) Antioxidation theory of non-steroidal anti-inflammatory drugs based upon the inhibition of luminol-enhanced chemiluminescence from the myeloperoxidase reaction. *Agents Actions* **12**: 371–376.

Pekoe, G. M., D. Peden, & K. Van Dyke (1983) Impairment of leukocyte myeloperoxidase bacterial mechanisms with ketamine (Ketalar®). *Agents and Actions* **13**: 59–62.

Pember, S. O., R. Shapira, & J. M. Kinkade, Jr. (1983) Multiple forms of myeloperoxidase from human neutrophilic granulocytes: Evidence for differences in compartmentalization, enzymatic activity, and subunit structure. *Arch. Biochem. Biophys.* **221**: 391–403.

Peters, R. A. & M. Shorthouse (1972) Fluorocitrate in plants and food stuffs. *Phytochem.* **11**: 1337–1338.

Petty, M. A. (1961) An introduction to the origin and biochemistry of microbial halometabolites. *Bact. Rev.* **25**: 111–130.

Phillips, K. A. & C. J. Rix (1981) Microprocessor-controlled determination of fluoride in environmental and biological samples by a method of standard additions with a fluoride ion selective electrode. *Anal. Chem.* **53**: 2141–2143.

Pickard, M. A. (1981) A defined growth medium for the production of chloroperoxidase by *Caldariomyces fumago. Can. J. Microbiol.* **27**: 1298–1305.

Pommier, J. & H. J. Cahnmann (1979) Interaction of lactoperoxidase with thiols and diiodotyrosine *J. Biol. Chem.* **254**: 3006–3010.

Purdy, M. A., J. Tenovuo, K. M. Pruitt, & W. E. White, Jr. (1983) Effect of growth phase and cell envelope structure on susceptibility of *Salmonella typhimurium* to the lactoperoxidase-thiocyanate-hydrogen peroxide system. *Infect. Immun.* **39**: 1187–1195.

Rabinowitz, J. L., M. Zanger, & V. Podolski (1976) Identification by nuclear magnetic resonance of iodinated lipids in the dog thyroid. *Biochem. Biophys. Res. Comm.* **68**: 1161–1168.

Ragan, M. A. (1981) Chemical constituents of seaweeds. *Botanical Monographs, (Oxford)* **17**: 589–626.

Ramakrishnan, K., M. E. Oppenhuizen, S. Saunders & J. Fisher (1983) Stereoselectivity of chloroperoxidase-dependent halogenation. *Biochem.* **22**: 3271–3277.

Rawitch, A. B., A. Taurog, S. B. Chernoff, & M. L. Dorris (1979) Hog thyroid peroxidase: Physical, chemical, and catalytic properties of the highly purified enzyme. *Arch. Biochem. Biophys.* **194**: 244–257.

Reed, H. L. (1980) The recovery of enzymes following their use in a reaction. U.K. Patent Office, Patent Application #GB 2,034,719A.

Reiter, B. & B. G. Harnulv (1982) The preservation of refrigerated and uncooled milk by its natural lactoperoxidase system. *Dairy Ind. Internat.* **47**: 13–19.

Renneberg, R., D. Pfeiffer, F. Scheller, & M. Janchen (1982) Enzyme sequence and competition electrodes based on immobilized glucose oxidase, peroxidase and catalase. *Analytica Chim. Acta* **134**: 359–364.

Repine, J. E., R. B. Fox, & E. M. Berger (1981) Dimethyl sulfoxide inhibits killing of *Staphylococcus aureus* by polymorphonuclear leukocytes. *Infect. Immun.* **31**: 510–513.

Rombauts, W. A., W. A. Schroeder, & M. Morrison (1967) Bovine lactoperoxidase. Partial characterization of the further purified protein. *Biochem.* **6**: 2965–2977.

Rose, A. F., J. A. Pettus, Jr., & J. J. Sims (1977) Marine natural products XIII. Isolation and synthesis of some halogenated ketones from the red seaweed *Delisea fimbriata. Tetrahedr. Letters* **22**: 1847–1850.

Rosen, H. & S. J. Klebanoff (1982) Oxidation of *Escherichia coli* iron centers by the myeloperoxidase-mediated microbial system. *J. Biol. Chem.* **257**: 13731–13735.

Rosenberg, I. N. & A. Goswami (1984) Iodotyrosine deiodinase from bovine thyroid. Ch. 31. In *Methods of Enzymology Volume 107 Posttranslational Modifications Part B*, F. Wold and K. Moldave (Ed.), Academic Press, New York, 488–500.

Sato, A. & W. Fenical (1983) Gramine-derived bromo-alkaloids from the marine bryozoan *Zoobotryon verticillatum. Tetrahedr. Letters* **24**: 481–484.

Saunders, B. C. (1957) Peroxidase. Action and Use in Organic Synthesis. In *The Royal Institute of Chemistry, Lectures, Monographs and Reports* No. 1, 1–27.

Schultz, J. (1980) Myeloperoxidase. *Reticuloendothel Syst.* **2**: 231–254.

Selvaraj, R. J., B. B. Paul, R. R. Strauss, A. A. Jacobs, & A. J. Sbarra (1974) Oxidative peptide cleavage and decarboxylation by the MPO-H_2O_2-Cl^- antimicrobial system. *Infect. Immun.* **9**: 255–260.

Selvaraj, R. J., J. M. Zgliczynski, B. B. Paul, & A. J. Sbarra (1978) Enhanced killing of myeloperoxidase-coated bacteria in the myeloperoxidase-H_2O_2-Cl^- system. *J. Infect. Dis.* **137**: 481–485.

Selvaraj, R. J., J. M. Zgliczynski, B. B. Paul, & A. J. Sbarra (1980) Chlorination of reduced nicotinamide adenine dinucleotides by myeloperoxidase: A novel bactericidal mechanism. *J. Reticuloendo. Soc.* **27**: 31–38.

Sergeev, G. B., A. I. Gudima, B. M. Sergeev, A. A. Taran, & V. A. Batyak (1981) Influence of peroxidase immobilization on kinetics of enzymatic oxidation of guaiacol in frozen solutions. *Biokhimiya* **46**: 986–989.

Sesin, D. F. & C. M. Ireland (1984) Iodinated phenethylamine products from a didemnid tunicate. *Tetrahedr. Letters* **25**: 403–404.

Shahangian, S. & L. P. Hager (1981) The reaction of chloroperoxidase with chlorite and chlorine dioxide. *J. Biol. Chem.* **256**: 6034–6040.

Shahangian, S. & L. P. Hager (1982) Isolation and characterization of horseradish peroxidase compound X. *J. Biol. Chem.* **257**: 11529–11533.

Shaw, P. D. & L. P. Hager (1959) An enzymatic chlorination reaction. *J. Amer. Chem. Soc.* **81**: 1011–1012.

Sheikh, Y. M. & C. Djerassi (1975) 2,6-Dibromophenol and 2,4,6-tribromophenols – antiseptic secondary metabolites of *Phoronopsis viridis. Experientia* **31**: 265–266.

Shimomura, K., M. G. Mullinix, T. Kakunaga, H. Fujiki, & T. Sugimura (1983) Bromine residue at hydrophilic region influences biological activity of aplysiatoxin, a tumor promoter. *Science* **222**: 1242–1244.

Siddiqi, I. W. (1982) An electrochemical assay system for peroxidase-couple reactions based on a fluoride ion-selective electrode. *Clin. Chem.* **28**: 1962–1967.

Siegel, B. Z. & A. W. Galston (1967) Indoleacetic acid oxidase activity of apoperoxidase. *Science* **157**: 1557–1559.

Sievers, G. (1979) The prosthetic group of milk lactoperoxidase is protoheme IX. *Biochim. Biophys. Acta* **579**: 181–190.

Sievers, G. (1981) Structure of milk lactoperoxidase. Evidence for a single polypeptide chain. *FEBS Letters* **127**: 253–256.

Sih, C. J. (1978) Glycosides of 2,6-bis (hydroxy-phenyl)-3,7-dioxabicyclo [3.3.0] octane. U.S. Patent Office, Patent #4,103,006.

Silverstein, R. M. & L. P. Hager (1974) The chloroperoxidase-catalyzed oxidation of thiols and disulfides to sulfenyl chlorides. *Biochem.* **13**: 5069–5073.

Sips, H. J. & M. N. Hamers (1981) Mechanism of the bactericidal action of myeloperoxidase: Increased permeability of the *Escherichia coli* cell envelope. *Infect. Immun.* **31**: 11–16.

Siuda, J. F. & J. F. De Bernardis (1973) Naturally occurring halogenated organic compounds. *Lloydia* **36**: 107–143.

Smal, E. & D. C. Baker (1984) 2′-Chloropentostatin (PD 115,243): Isolation, identification and synthesis of the sugar moiety. Abstract #8, Division of Carbohydrate Chemistry, 188th ACS National Meeting.

Smith, A. M., W. L. Morrison & P. J. Milham (1983) Ethylene enhances reactivity of superoxide with peroxidase to form the oxy-ferrous complex. *Biochem.* **22**: 1645–1650.

Smith, D. C. & S. J. Klebanoff (1970) A uterine fluid-mediated sperm-inhibitory system. *Biol. Reprod.* **3**: 229–235.

Smith, Q. T. & C. H. Yang (1984) Salivary myeloperoxidase of young adult humans (41822). *Proc. Soc. Experiment. Biol. Medicine* **175**: 468–475.

Soiefer, A. I. & P. J. Kostyniak (1984) Purification of a fluoroacetate – specific defluorinase from mouse liver cytosol. *J. Biol. Chem.* **259**: 10787–10792.

Starr, M. P., C. L. Jenkins, L. B. Bussey, & A. G. Andrews (1977) Chemotaxonomic significance of the xanthomonadins, novel brominated aryl-polyene pigments produced by bacteria of the genus *Xanthomonas. Arch. Microbiol.* **113**: 1–9.

Stelmaszynska, T. & J. M. Zgliczynski (1963) Oxidation of cystine by hydrogen peroxide catalyed by horseradish peroxidase. *Acta Biochim. Polonica* **10**: 371–378.

Stijve, T. (1984) Inorganic bromide in higher fungi. *Z. Naturforsch* **39c**: 863–866.

Suflita, J. M., A. Horowitz, D. R. Shelton, & J. M. Tiedje (1982) Dehalogenation: A novel pathway for the anaerobic biodegradation of haloaromatic compounds. *Science* **218**: 1115–1117.

Sun, H. H. & W. Fenical (1979) Rhipocephalin and rhipocephenal; toxic feeding deterrents from the tropical marine alga *Rhipocephalus phoenix. Tetrahedr. Letters* **8**: 685–688.

Sun, H. H., V. J. Paul, & W. Fenical (1983) Avrainvilleol, a brominated diphenylmethane derivative with feeding deterrent properties from the tropical green alga *Avainvillea longicaulis. Phytochem.* **22**: 743–745.

Takahashi, K., H. Nishimura, T. Yoshimoto, Y. Saito, & Y. Inada (1984) A chemical modification to make horseradish peroxidase soluble and active in benzene. *Biochem Biophys. Res. Comm.* **121**: 261–265.

Takala, K., P. Kauranen, & R. Fagersten (1979) Fluorine content of terricolous lichens and bryophytes on exposed rapakivi bedrock. *Ann. Bot. Fennici* **16**: 90–92.

Tamura, M., N. Shimidzu, & K. Hayashi (1977) Vanadium-substituted hemoproteins (I). *Biochem. Biophys. Res. Comm.* **75**: 1029–1036.

Tanaka, N., T. Satake, A. Takahasihi, M. Mochizuki, T. Murakami, Y. Saiki, J. Z. Yang, & C. M. Chen (1982) Chemical and chemotaxonomical studies of ferns. XXXIX. Chemical studies on the constituents of *Pteris bella* Tagawa and *Pteridium aquilinum* subsp. *wightianum* (Wall) Shich. *Chem. Pharm. Bull.* **30**: 3640–3646.

Tenovuo, J., J. Valtakoski, & M. L. E. Knuuttila (1977) Antibacterial activity of lactoperoxidase adsorbed by human salivary sediment and hydroxyapatite. *Caries Res.* **11**: 257–262.

Tenovuo, J. & K. Kurkijarvi (1981) Immobilized lactoperoxidase as a biologically active and stable form of an antimicrobial enzyme. *Archs Oral Biol.* **26**: 309–314.

Tenovuo, J., K. M. Pruitt, & E. L. Thomas (1982) Peroxidase antimicrobial system of human saliva: Hypothiocyanite levels in resting and stimulated saliva. *J. Dent. Res.* **61**: 982–985.

Tenovuo, J., K. K. Makinen, & G. Sievers (1985) Antibacterial effect of lactoperoxidase and myeloperoxidase against *Bacillus cereus. Antimicrob. Agents Chemother.* 27, 96–101.

Theiler, R. F., J. S. Siuda, & L. P. Hager (1978a) Bromoperoxidase from the red algae *Bonnemasisonia hamifera.* In *Drugs and Food from the Sea. Myth or Reality*, P. N. Kaul and C. J. Sindermann, (Eds.), Univ. Of Oklahoma Press, 153–169.

Theiler, R., J. C. Cook, L. P. Hager, & J. F. Siuda (1978b) Halohydrocarbon synthesis by bromoperoxidase. *Science* **202**: 1094–1096.

Thibault, P., P. Monsan, & C. Jouret (1981) Influence of ethanol on the activity and the stability of free and immobilized peroxidase. *Sciences des Aliments* **1**: 55–65.

Thomas, E. L. & T. M. Aune (1977) Peroxidase-catalyzed oxidation of protein sulfhydryls mediated by iodine. *Biochem.* **16**: 3581–3586.

Thomas, E. L. & T. M. Aune (1978a) Susceptibility of *Escherichia coli* to bactericidal action of lactoperoxidase, peroxide, and iodide or thiocyanate. *Antimicrob. Agents Chemother.* **13**: 261–265.

Thomas, E. L. & T. M. Aune (1978b) Oxidation of *Escherichia coli* sulfhydryl components by the peroxidase-hydrogen peroxide-iodide antimicrobial system. *Antimicrob. Agents Chemother.* **13**: 1006–1010.

Thomas, E. L., K. P. Bates, & M. M. Jefferson (1981) Peroxidase antimicrobial system of human saliva: Requirements for accumulation of hypothiocyanite. *J. Dent. Res.* **60**: 785–796.

Thomas, E. L., M. M. Jefferson, & M. B. Grisham (1982) Myeloperoxidase-catalyzed incorporation of amines into proteins: Role of hypochlorous acid and dichloramines. *Biochem.* **21**: 6299–6308.

Thomas, E. L., M. B. Grisham, & M. M. Jefferson (1983a) Myeloperoxidase-dependent effect of amines on functions of isolated neutrophils. *J. Clin. Invest.* **72**: 441–454.

Thomas, E. L., K. A. Pera, K. W. Smith, & A. K. Chwang (1983b) Inhibition of *Streptococcus mutans* by the lactoperoxidase antimicrobial system. *Infect. Immun.* **39**: 767–778.

Thomas, J. A. (1968) Studies on the mechanism of the halogenation reactions catalyzed by chloroperoxidase. Ph.D. Thesis, University of Illinois, 146 pp.

Thomas, J. A. & L. P. Hager (1968) The peroxidation of molecular iodine to iodate by chloroperoxidase. *Biochem. Biophys. Res. Comm.* **32**: 770–775.

Thomas, J. A., D. R. Morris, & L. P. Hager (1970) Chloroperoxidases. VII Classical peroxidatic, catalytic, and halogenating forms of the enzyme. *J. Biol. Chem.* **245**: 3129–3131.

Thomas, S. O., J. A. Lowery, & V. L. R. Singleton (1959) Nucleocidin and the process of obtaining the same. U.S. Patent Office, Patent #2,914,525.

Thomson, R. H. (1978) Halogenated metabolites from marine animals and plants. *J. Indian Chem. Soc.* **60**: 1209–1215.

Tinsley, I. J. & R. R. Lowry (1980) Bromine content of lipids of marine organisms. *J. Amer. Oil Chem. Soc.* *57*: 31–33.

Travis, J. K. Beatty, & N. Matheson (1984) Oxidation of alpha-1-proteinase inhibitor: Significance for pathobiology. *Adv. Experiment Medical Biol.* **167**: 89–95.

Tsan, M.-F. and J. W. Chen (1980) Oxidation of methionine by human polymorphonuclear leukocytes. *J. Clin. Invest.* **65**: 1041–1050.

Tsan, M.-F. (1982a) Myeloperoxidase-mediated oxidation of methionine and amino acid decarboxylation. *Infect. Immun.* **36**: 136–141.

Tsan, M.-F. (1982b) Myeloperoxidase-mediated oxidation of methionine. *J. Cell. Physiol.* **111**: 49–54.

Turk, J., W. R. Henderson, S. J. Klebanoff, & W. C. Hubbard (1983) Iodination of arachidonic acid mediated by eosinophil peroxidase, myeloperoxidase and lactoperoxidase. Identification and comparison of products. *Biochem. Biophys. Acta* **751**: 189–200.

Turkall, R. M. & M.-F. Tsan (1982) Oxidation of glutathione by the myeloperoxidase system. *J. Reticulendo. Soc.* **31**: 353–360.

Turner, W. B. (1971) *Fungal Metabolites.* Academic Press, New York.

Turner, W. B. & D. C. Aldridge (1983) *Fungal Metabolites II.* Academic Press, New York.

van Pee, K.-H., O. Salcher, P. Fischer, M. Bokel & F. Lingens (1983) The biosynthesis of brominated pyrrolnitin derivatives by *Pseudomonas aureofaciens. J. Antibiot.* **36**: 1735–1742.

van Pee, K.-H. & F. Lingens (1984) Detection of a bromoperoxidase in *Streptomyces phaeochromogenes. FEBS* **173**: 5–8.

van Pee, K.-H. & F. Lingens (1985) Purification of a bromoperoxidase from *Pseudomonas aureofaciens. J. Bacteriol.* **161**: 1171–1175.

Van Zyl, A., B. Burger, P. Rossouw & B. Wilson (1981) The control of peroxidase-catalyzed iodination and de-iodination. *S. Afr. Med. J.* **60**: 184–191.

Vartiainen, T. & P. Kauranen (1984) The determination of traces of fluoroacetic acid by extractive alkylation pentafluorobenzylation and capillary gas chromotography-mass spectrometry. *Anal. Chim. Acta* **157**: 91–97.

Vickery, B., M. L. Vickery, & F. Kaberia (1979) A possible biomimetic synthesis of fluoroacetic acid. *Experimentia* **35**: 299–300.

Vilter, H. (1983) Peroxidases from Phaeophyceae. IV. Fractionation and location of peroxidase isoenzymes in *Ascophyllum nodosum* (L.) Le Jol. *Bot. Marina* **26**: 451–455.

Vilter, H., K.-W. Glombitza, & A. Grawe (1983) Peroxidases from Phaeophyceae. I: Extraction and detection of the peroxidases. *Bot. Marina* **26**: 331–340.

Vilter, H. (1984) Peroxidases from Phaeophyceae: A vanadium (V)-dependent peroxidase from *Ascophyllum nodosum. Phytochem.* **23**: 1387–1390.

Walker, J. R. L. & B. C. Lien (1981) Metabolism of fluoroacetate by a soil *Pseudomonas* sp. and *Fusarium solani. Soil. Biol. Biochem.* **13**: 231–235.

Walker, R. P. & D. J. Faulkner (1981) Chlorinated acetylenes from the nudibranch *Diaulula sandiegensis. J. Org. Chem.* **46**: 1475–1478.

Wang, M.-Y., B. M. Hoffman, & P. F. Hollenberg (1977) Cobalt-substituted horseradish peroxidase. *J. Biol. Chem.* **252**: 6268–6275.

Ward, P. F. V., R. J. Hall, & R. A. Peters (1964) Fluoro-fatty acids in the seeds of *Dichapetalum toxicarium. Nature* **201**: 611–612.

Weiss, S. J., R. Klein, A. Slivka, & M. Wei (1982) Chlorination of taurine by human neutrophils. Evidence for hypochlorous acid generation. *J. Clin. Invest.* **70**: 598–607.

Weiss, S. J., M. B. Lampert, & S. T. Test (1983) Long-lived oxidants generated by human neutrophils: Characterization and bioactivity. *Science* **222**: 625–627.

Weiss, S. J., G. Peppin, X. Ortiz, C. Ragsdale, & S. T. Test (1985) Oxidative autoactivation of latent collagenase by human neutrophils. Science **227**: 747–749.

Weitzman, S. A., A. B. Weitberg, E. P. Clark, & T. P. Stossel (1985) Phagocytes as carcinogens: Malignant transformation produced by human neutrophils. *Science* **227**: 1231–1233.

Welinder, K. G. & L. B. Smillie (1972) Amino acid sequence studies of horseradish peroxidase. II. Thermolytic peptides. *Can. J. Biochem.* **50**: 63–90.

Welinder, K. G., L. B. Smillie, & G. R. Schonbaum (1972) Amino acid sequence studies of horseradish peroxidase. I. Tryptic peptides. *Can. J. Biochem.* **50**: 44–62.

Welinder, K. G., (1976) Covalent structure of the glycoprotein horseradish peroxidase (EC 1.11.1.7). *FEBS Letters* **72**: 19–23.

Wever, R., M. N. Hamers, R. S. Weening, & D. Roos (1980) Characterization of the peroxidase in human eosinophils. *Eur. J. Biochem.* **108**: 491–495.

Wever, R., H. Plat, & M. N. Hamers (1981) Human eosinophil peroxidase: A novel isolation procedure, spectral properties and chlorinating activity. *FEBS Letters* **123**: 327–331.

White, R. H. & L. P. Hager (1977) Occurrence of fatty acid chlorohydrins in jellyfish lipids. *Biochem.* **16**: 4944–4948.

White, W. W., Jr., K. M. Pruitt, & B. Mansson-Rahemtulla (1983) Peroxidase-thiocyanate-peroxide antibacterial system does not damage DNA. *Antimicrob. Agents Chemother.* **23**: 267–272.

Whyte, J. N. C. & J. R. Englar (1976) Determination of halogens in marine algae by use of an ion-selective electrode. *Analyst* **101**: 815–819.

Wijekoon, W. M., E. Aynoglu, & C. Djerassi (1984) Phospholipid studies of marine organisms 9. New brominated demospongic acids from the phospholipids of two *Petrosia* species. *Tetrahedr. Letters* **25**: 3285–3288.

Wolk, C. P. (1968) Role of bromine in the formation of the refractile inclusions of the vesicle cells of the *Bonnemaisoniaceae* (Rhodophyta). *Planta* **78**: 371–378.

Wood, W. F. (1983) Chemical ecology: Chemical communication in nature. *J. Chem. Educ.* **60**: 531–539.

Wright, C. D. & R. D. Nelson (1985) Candidacidal activity of myeloperoxidase: Therapeutic influence of the enzyme *in vivo*. *Infect. Immun.* **47**: 363–365.

Wulff, P., J. S. Carle, & C. Christophersen (1982) Marine alkaloids--6. The first naturally occurring bromo-substituted quinoline from *Flustra foliacea*. *Comp. Biochem. Physiol.* **71B**: 525–526.

Yamada, M., M. Mori, & T. Sugimura (1981) Purification and characterization of small molecular weight myeloperoxidase from human promyelocytic leukemia HL-60 cells. *Biochem.* **20**: 766–771.

Yamada, M. (1982) Myeloperoxidase precursors in human myeloid leukemia HL-60 cells. *J. Biol. Chem.* **257**: 5980–5982.

Yamazaki, I., M. Tamura, Y. Kaneko, & Y. Kuwahara (1982) The oxidation of Zn- and Mg-substituted horseradish peroxidases. In *Oxygenases and Oxygen Metabolism*, M. Nozaki, S. Yamamoto, Y. Ishimura, M. J. Coon, L. Emster, & R. W. Estabrook, Eds., Academic Press, New York, 457–465.

Yonetani, T. & T. Asakura (1969) Studies on cytochrome c peroxidase. XV. Comparison of manganese porphyrin-containing cytochrome c peroxidase, horseradish peroxidase, and myoglobin. *J. Biol. Chem.* **244**: 4580–4588.

Yoshida, M., T. Frjita, N. Kurihara, & M. Nakajima (1985) Dehydrochlorination of p,p′-DDT by cultured cells derived from ovarian tissues of *Culex pipiens* var. molestus. *Pest. Biochem. Biophysiol.* **23**: 1–6.

Zajak, M., L. Bjorck & O. Claesson (1981) Antibacterial effect of the lactoperoxidase system sgainst *Bacillus cereus*. *Milchwissenschaft* **36**: 417–418.

Zgliczynski, J. M., T. Stelmaszynska, E. Olszowska, A. Krawczyk, E. Kwasnowska, & J. T. Wrobel (1983) Peroxidative oxidation of halides catalysed by myeloperoxidase. Effect of fluoride on halide oxidation. *Acta Biochim. Polonica* **30**: 213–222.

Index